SPIDER EVOLUTION

SPIDER EVOLUTION

Genetics, Behavior, and Ecological Influences

SUBIR RANJAN KUNDU

Edited by

JAGMEET KAUR

ELSEVIER

ACADEMIC PRESS

An imprint of Elsevier

Academic Press is an imprint of Elsevier
125 London Wall, London EC2Y 5AS, United Kingdom
525 B Street, Suite 1650, San Diego, CA 92101, United States
50 Hampshire Street, 5th Floor, Cambridge, MA 02139, United States
The Boulevard, Langford Lane, Kidlington, Oxford OX5 1GB, United Kingdom

Library of Congress Cataloging-in-Publication Data
A catalog record for this book is available from the Library of Congress

British Library Cataloguing-in-Publication Data
A catalogue record for this book is available from the British Library

ISBN: 978-0-323-90041-6

For information on all Academic Press publications visit our website at https://www.elsevier.com/books-and-journals

Publisher: Charlotte Cockle
Acquisitions Editor: Anna Valutkevich
Editorial Project Manager: Andrea Dulberger
Production Project Manager: Swapna Srinivasan
Cover Designer: Christian J. Bilbow

Typeset by TNQ Technologies

To Anna Valutkevich—
With gratitude to my acquisitions editor for inspiring me to explore the uncharted territory of "evolutionary biology."

Contents

Foreword

I am honored to write the foreword for this book, nicely written by Subir Ranjan Kundu. From my background in conservation and my personal objectives to address all conflicts with any form of wildlife, from ants to elephants, I know spiders to be a group of creatures maligned by humanity. Because of nonsensical ignorance, humans are often unnecessarily mystified by spiders. People exaggerate that a spider bite from this or that spider has become necrotic and causes severe pain. In fact, the person bitten may not have even seen the spider, and in any event, may not be able to identify the exact spider species. Accusations about and identifications of spiders as Species X or Species Y are often made in ignorance.

I therefore urge all readers to consider Systems Thinking—that is, cause and effect—because when one questions cause and effect, many situations become clearer and easily understandable. Spiders may bite humans when they feel threatened, from fear of being crushed by a sleeping human, or when trapped within folds of clothing. Therefore, human avoidance of such situations is a way to minimize the risk of spider bites. When considering envenomation, bear in mind that all forms of envenomation or poisoning are dosage dependent, and very few spiders have venom that is potentially harmful to humans. Furthermore, their mouthparts are adapted for capturing small prey and feeding on that small prey; as such, even if spider venom were toxic to humanity, the spider's mouthparts are unable to administer an effective envenomation bite.

Spiders have the beneficial role of catching and consuming flies and other insects. By greater acceptance of spiders, humanity can use fewer pesticides and insecticides, making our own environments healthier. Spiders may be trapped for removal from our homes by using a glass jar to enclose them and then sliding a sheet of paper between the surface on which they are standing and the glass jar top, thus enclosing them so they can be carried and deposited outdoors where they can continue their existence. Human conflicts with wildlife cannot and should not be addressed by killing creatures, as this merely offers a short-term solution without addressing the issue holistically and logically.

I hope readers will reap benefits from this book and learn to respect and cherish all forms of life on our planet by learning, understanding, and respecting the complexities of all life around us.

Timothy Snow
10 December 2020
Wildlife Poisoning Prevention & Conflict Resolution,
Rosetta, KwaZulu-Natal, South Africa
E-mail: snowman@bundunet.com; Skype: timsnow.snowman;
Website: www.wildlifepoisoningprevention.co.za

Acknowledgments

I express my deep sense of gratitude and indebtedness to Dr. Ayyanadar Arunachalam, Assistant Director-General, Indian Council of Agricultural Research, New Delhi, India; Ms. Lata Sardesai, Animal Physiologist, Head—Department of Zoology, R.D. National and S.W.A. Science College, Bandra, Mumbai, India; and Dr. Pradipta Kumar Sarangi, Lecturer in Zoology, Indira Memorial College, Chandiput, Gajapati, Odisha, India, for their suggestions and guidance in this present work.

I am especially thankful to Mr. Timothy Snow, Wildlife Poisoning, Prevention & Conflict Resolution, Rosetta; Dr. P. U. Antony, Professor of Zoology, Christ University, Bangalore, India; and Dr. Sainudeen Pattazhy, Associate Professor, Department of Zoology, University of Kerala, Trivandrum, India, who helped me draft various parts of the manuscript, reviewed it, provided consultation on literature and library research, engaged in discussion, etc.

I express my gratitude to Sung-Min Han and Susan Zadek, Albert Campbell Library, and Eva Lew, Bloor Gladstone Library, Toronto Public Library Network, Toronto, Canada, for extending their library facilities to me and providing necessary help in locating the literature required to complete this project.

I am also thankful to Mr. Goutam Saha, the freelance graphics artist from Kolkata, India, for preparing the graphical presentations in this book.

I am thankful to all academicians, editors, publishers of books and journals, and colleagues for contributing and offering printed and digital materials, tendering valuable suggestions, and sharing necessary clarifications over the years while they selflessly helped me in getting this task done.

CHAPTER ONE

Introduction: A fable of the first fire, brought by the water spider from a warm glowing island to the cold dark world of animal-people

Abstract

The spider is regarded as a symbol of perseverance for its hunting sprees to entrap prey by spinning webs. The spinning potential of spiders as culturally depicted in the origins of weaving nets and spinning textiles, and integrated in creation myths since the dawn of civilization, has currently been attenuated to the metaphoric invisible connectivity in the world of the "World Wide Web." In the history of human civilization of different parts of the world (from Greco—Roman mythologies to Afro—Asian folktales), the spider has appeared as a symbolic spiritual entity, mostly as a symbol of ill fate because of its prey-capturing potential using venom that causes slow death of the prey. Since their origin, spiders have mostly occupied the center stage of diverse sociocultural histories and the folklore of human civilizations as a symbol of misfortune, detriment, animosity, and fear mainly due to our biased perceptions of this creature.

Keywords: Folklore; Mythology; Spider; Symbolic entity; Weaving web.

Humanity is an enormous spider web, so that if you touch it anywhere, you set the whole thing trembling.... Our lives are linked together. No man is an island.

Frederick Buechner

Apparently, to the majority of people in our socioeconomic community in this fast-forward life, there is little or no importance to nurturing inconsequent fragments of memories like the obsolete word "alms" (in the form of clothing, foods, etc.), used in CE 1800 to narrate symbolic charity or donation to the poor and the Bohemian travelers trotting from one place to another. Though the primary objective of exploring a new place is fulfilled by both categories of people, tourists and travelers alike, there is a clear distinction between the two. A tourist explores and enjoys the experience of

Spider Evolution
ISBN: 978-0-323-90041-6
https://doi.org/10.1016/B978-0-323-90041-6.00011-5

traveling/touring without compromising on standards of living, existing lifestyle, food habits, and cultural and socioeconomic niches, while a traveler always strives to experience traveling by trying to blend in with the native cultures, tasting new types of foods, and grasping a unique type of exposure by exploring beyond their socioeconomic stereotypes. Eventually, the blending-in experiences of travelers, which were mostly considered inconsequent compared with those of others, might render the travelers lost outside the cartographic setup but drag them along to enjoy the adventurous experience of seeing or witnessing a strange, beautiful, or sublime fragment of life—the ultimate sublime, serendipitous alms of the traveler.

What follows is about one of my finest travel experiences, which happened around 9 years ago on August 21, 2019, when I traveled from Toronto to Oakville—from one of the busiest metropolises in southern Canada to one of the busy towns in the Greater Toronto Area (GTA), located just 35 km from Toronto proper. I was quite thrilled to share with my colleagues at work that I was going to visit my erstwhile neighbor and friend, Ronen Goldstein, and his family to help them design and install a water garden in the backyard of their newly bought duplex in Oakville, but nearly everyone stared at me with uncomfortable looks, probably thinking there was nothing special about this since Oakville is a crowded town in the GTA (it's not a tourist spot like the North Bay or Georgian Bay or Algonquin Park) and it's only about a 40 min commute from the Coach Canada terminal downtown to Oakville Centre (near the Holiday Inn on Argus Road). Perhaps I had not expressed my state of mind to my friends and colleagues correctly, that Ronen was my next-door neighbor at some point in time and that we had lived in the same neighborhood for 3 years, and it always feels great spending time with your old friends, though there's nothing special in spending some time with old neighbors. It was going to be a unique experience for me as maintaining a formal relationship with a neighbor stops anyone from getting into their sociocultural ambiance and being a part of it even for a short period. It was quite intriguing to me to get in touch with the composite cultural milieu of a wonderful couple, Ronen and Genevieve, who came from completely two different ethnicities, religious and sociocultural backgrounds. A prime attraction was their cute 7-year-old daughter Eleanor, an adorable, calm, and quiet kid whom I consider one of the best storytellers in the Bloor/Gladstone Library of my erstwhile neighborhood, Bloordale Village, at the west end of Toronto.

I must say something about the ethnic backgrounds of Genevieve and Ronen; they blossomed into a nice couple despite coming from different ethnoreligious backgrounds. Ronen was raised in an orthodox Jewish family; his ancestors settled in Toronto during the Second World War to escape religious persecution in Germany. His grandfather was a well-known Rabbi in the Jewish community of Toronto, particularly in the Bathurst area.

On the other hand, Genevieve was born on the Eastern Cherokee Reserve and spent her childhood with her parents among the members of the Eastern Band of Cherokee Indians, who had settled at the outskirts of the Nantahala National Forest in North Carolina, USA. After her parents separated, she lost contact with them and the other members of the Band. She came to Toronto, Canada, with her uncle at the age of 12, and since then she had been an active member of the Evangelical Church of America and had also remained a bona fide Torontonian.

I was lucky that I was with Ronen in the Mount Sinai Hospital, Downtown Toronto, when Eleanor was born on a cold freezing night in December. In the final stage of pregnancy, Genevieve was getting sick very often, so we were very worried about the health of the mother and the would-be child. I was anxiously waiting in the downstairs lobby when Ronen broke the good news to me. I was so happy to see their daughter, our little butterfly, Eleanor. Initially, Eleanor was a little underweight and that made Ronen scared to hold her tightly in his lap and so he laid her down on my lap. Within a couple of days of her birth, Eleanor made her first appearance in her home with her father and mother. She came home from the hospital on my lap, and even now Ronen and Genevieve tease me about this—all through the ride inside the car, I cried in joy. I could never explain why I did that, but it was one of the best, most cherished moments of my life. It was wonderful to hold a newborn abreast and feel her heartbeat, and it would remain in my memory forever. Whatever the reason, I always feel that Eleanor is a special gift to me, a bundle of happiness that can cure all my situations and help me get through difficult situations in life. When Genevieve broke the news that they planned to relocate to Oakville, I was almost dumbstruck; I could just manage to ask her the exact date when they planned to move, and I booked a ticket for 3 weeks' vacation to Calcutta so that I didn't have to say goodbye to my little butterfly. After coming back from my vacation, I came to know that our beloved butterfly asked her father about my whereabouts, and she cried for 2 to 3 months after they moved to Oakville from Toronto. But I was not emotionally strong enough to face her at the time, so I avoided visiting their place for a long while. I did not even avail myself of a vacation for 2 years; it was difficult for me to stay at home, especially during the weekends, without the chatter of our little butterfly or our routine visits to her favorite coffee shop, Tim Hortons, at the other side of the Bloor/Gladstone Library. Literally, I felt like I was stuck somewhere in the middle of a long tunnel of seclusion where silence grasped me in such a way that all colors around me were faded out. So after a couple years of working at a stretch without taking a single day off, when I approached the human resources manager of my organization to request leave for August 21, 2009 (Friday), and told her that the reason for the leave was to help my family friend who lived in Oakville design and build a water garden in his backyard, she was a little confused by the reason I gave and was unable to justify whether I had given the right reason or was just bluffing, but she gave me a twisted smile and approved it immediately.

It was quite thrilling for me to stay with my friends, and not in a hotel environment but in a new place with a totally different culture and ethnicity, which is when it comes to me that I am wholly, solely multiculturalist. We had planned earlier that we would finish Eleanor's "Water Garden Installation" project within 48 h, from Friday to Saturday, so we could hang out on Sunday before my return to my place in Toronto.

On Friday around 11:30 a.m., I reached the Toronto Coach Terminal (Bay Street) to catch the Megabus scheduled to leave around half past 12; I arrived at the Holiday Inn, Oakville (Argus Road), in around 40 min. I was aware that on my arrival Ronen would come to receive me, but it was like heaven's surprise that he arrived with his whole family. When I saw Genevieve (Ronen's wife) and little Eleanor, my heart bounced with happiness, as I myself didn't know how much I had missed her company and her unique stories (with a touch of magic realism), which she used to tell me in those idle weekends afternoons when she was my next-door neighbor. As far as I can remember, my vision became blurred; it was difficult to see through my foggy glasses, but I still managed to get a glimpse of the beaming face of our beloved little butterfly Eleanor as she ran toward me. After a round of bear hugs, we rushed to Tim Hortons near the Oakville Center of Trafalgar Road for my favorite coffee. When Ronen volunteered to order a big box of Timbits with French vanilla for Eleanor and coffees for us, he stared at me to find out how I like my coffee, and I noticed, surprisingly, that Eleanor mentioned from the side, "It's a large, double double[1]

[1] large double double: It's a popular form of ordering coffee in a large size of mug or paper cup in any coffee shop in North America with regular blend coffee with two spoon of milk and two spoon of sugar in it.

and a butter croissant for Subi." I was feeling a little awkward because I didn't want Ronen to get a croissant for me, but before I could stop him, he ordered, and I just looked at Eleanor and realized again that it does not matter what we consider our home, what matters is what we feel about "our home."

Home might be a spot on a cartographic chart, but that does not make anyone feel that they belong to that place. I was suddenly plunged into the memories of those nice stories that I had heard from little Eleanor. She once told me a story about "Thunderbird," the mythological legendary bird of North American aboriginal communities that does not have any specific home, and wherever it stops, that place becomes its settlement. I started feeling that Eleanor had brought me Home; earlier, I did not know what place I belonged to. She still remembered that when I took her (her parents were busy even on the weekends, as both were paramedics) to Tim Hortons coffee shop beside the Bloor/Gladstone Library after attending her monthly children's literary workshop, I would order a "large double double" along with a butter croissant for myself, and for her, medium-sized French vanilla with a small box of Timbits.

At the coffee table, Ronen and Genevieve asked me about their old neighborhood and the health of a number of retired people, as we always met at the neighborhood programs. Our cute butterfly was watching us and waiting patiently, as she had not gotten a chance to tell me about her school, friends, library workshops, etc. After an hour, she reminded us that we were supposed to pick up utilities, accessories, building materials, lumber, some water plants, colored fishes, and so on for a project and that we needed to do the prep work for the next day. So immediately, we left the coffee shop and got into the station wagon to go to Home Depot at the corner of Cross Avenue to shop, and we ended up buying everything on the shopping list. We bought about a 100 pieces of paving stones, a few bags of crushed limestones, ready-mix concrete, a big PVC pond liner, some pressure-treated lumber, etc., and loaded the trailer attached behind the station wagon, then picked up pond plants (American lotus) from a local garden nursery and a number of fishes from a local pet store at Trafalgar Road and went straight home to dump everything in the backyard except for the live goods. Immediately after getting home, Ronen, Genevieve, and Eleanor got dressed for our little outdoor project, and Ronen and I took the lead in the excavation and creation of a base for installing the pond-liner, while Genevieve and Eleanor put all the garden waste, dry grass, broken branches, and fallen leaves into garden waste bags and placed them in the green trash bins at the corner of the backyard. After doing the prep work to install the pond liner on the base, we finished and decorated it by laying the cobblestone around it and connecting the plumbing line of the water fountain to the pond. After 3 to 4 h of effort and hard work, Ronen and Genevieve left to get dinner ready while Eleanor decided to stay with me to finish cleaning all the debris so we could start early the next day and get the job done before the middle of the day.

I can still recall that it was getting dark that evening and that Eleanor kept insisting that we go inside so she could show me all her new storybooks. She also wanted to share her new collection of stories, ones she had heard and learned from Genevieve. I noticed some weeds and dandelions at the bottom of a tree, and I thought I would clean it up before taking a shower, going to dinner, and attending her storytelling session. Dandelions are difficult to pull up if their rootstock is entangled with the rootstock of a nearby bush. It was difficult for me to pull these dandelions up with garden gloves on, so I took them off and proceeded to uproot the weeds. As I was uprooting the weeds, I suddenly felt severe pain in the index finger of my right hand; it seemed that a spiderlike insect had stung me. There was no sign of a prick on the surface of the skin, but a tiny drop of blood oozed out of a tiny opening somewhere. It struck me then that a venomous creature might have bitten me, probably a poisonous snake or spider.

Immediately after experiencing the severe pain, my finger started burning, and I could not see properly anymore as everything became blurred and I struggled to shout out a name—"Kamama .... Kamama." I tried to say something about my index finger but failed to communicate properly, and I collapsed in the backyard. Because of that venomous sting, my brain was in a semianemic anesthetic stage, but then I heard a faint, faint noise, someone repeatedly saying, "Ama ....Pa Ama ....Pa ... Akkabish[2] stung Subi." I was not fully unconscious, as I could hear the faint sounds, but I could not see anything properly, and I found myself helplessly lying in the backyard garden drenched in sweat. I failed to gather the strength to sit up. I felt Eleanor sobbing softly beside me while her parents tried to figure out what had happened to me before she called 911 for emergency assistance; then, the ambulance was on its way to attend the emergency. In the meantime, I heard that Eleanor had noticed that it was a medium-sized black spider

[2] Akkabish: It is a Biblical word means "Spider," which has been used in Bible for few occasions (Ref. Isaiah 59:5, Biblical Old Testament).

that had stung me when I was trying to uproot the dandelions beside the rootstock of the tree.

Being paramedics, Ronen and Genevieve were very much aware that there was an occurrence of the black widow spider in the area, and just recently, an older lady, one of Ronen's colleagues, was admitted to the Oakville–Trafalgar Memorial Hospital, and the news had appeared in all the leading newspapers, such as the Toronto Star (Reiti, 2009). Ronen and Genevieve were very tense and kept a sharp eye on me to see if I showed any typical symptoms of venom-induced ailments. My body fell into a state of numbness, particularly its lower half. The paramedics recorded everything and took me to the Oakville–Trafalgar Memorial Hospital for treatment and observation. I was in a state of shock; the pain and burning sensations of the spider stings were blinding. There were other physical symptoms like blurred vision, numbness, etc. I could not notice and gauge the mental states of Ronen and Genevieve; I could just feel our butterfly sobbing, which made me want to cry out loudly, but all I could utter was "Kamama ....everything will be alright."

After being admitted to the Trafalgar Memorial Hospital, I felt anxious and helpless until Ronen and Genevieve came to see me. They mentioned that our "Kamama" was a little upset about my physical condition and was waiting in the lobby to find out how I was.

Even in my half-conscious condition, I recalled the old saying "What goes around, comes around." Under the influence of medication, I felt a little dizzy, and a frozen shot appeared in my subconscious mind of a soft-towel wrapped around a little babe in my lap, and I came out of Mount Sinai Hospital to get into a car with a unique feeling of happiness and slowly plunged into a half-sleeping state.

During the summers, whenever Genevieve had a day off, she spent some time in the park with little Eleanor and often requested that I join them, and I was enthralled by Eleanor's childhood stories from her time on the Cherokee Indian Reserve; the stories sounded like fairytales from a different world, a world we were not familiar with, a world of mountain ranges where the Moon and the Sun had an identity, where gods prevailed and the spirits of animals dwelled among the humans.

At that time, she was recognized by the members of her Band by a different name, "Ama," which means water; her parents had chosen this name because she was born after Hurricane Camille flooded the west part of North Carolina in 1969.

Hurricanes are tropical storms known to be caused by the Mayan god Huracan, or Taino. In Caribbean–Indian folklore, the god of evil, "Huraken" or "Jurakan" (Emanuel, 2005; Saunders and Lea, 2019; Oblack, 2020), was responsible for these hurricanes.

Before getting into the mythological history of hurricanes, it should be clarified here that hurricanes developed when tropical cyclones spun around in the low-pressure zones (64 knots) of the western hemisphere areas of the Caribbean Sea, Gulf of Mexico, North Atlantic Ocean, and Eastern and Central North Pacific Ocean. It should be mentioned that in 1848, Mr. Henry Piddington, the erstwhile curator of the Indian Museum, had used the term "cyclone" for the first time, and in Greek, the meaning of this word was "coil of a snake" which is quite descriptive of the storms created by rotating winds in the low-pressure areas of tropical oceans.

It should be made clear that all hurricanes are tropical cyclones, but not all tropical cyclones are hurricanes. Cyclones with a velocity lower than 39 mph are recognized as depressions; those falling in a velocity range of 39–73 mph are recognized as cyclones; and cyclonic storms above a velocity of 74 mph are recognized as hurricanes. It has also been observed by meteorologists that the center of the hurricane always remains calm. However, its eye wall is the most vulnerable zone, and the eye of the hurricane ranges from 5 to 120 miles in diameter (Avila and Cangialosi, 2011).

Genevieve, aka Ama, had spent her childhood on the Cherokee Indian Reserve in North Carolina, USA, and had dreadful memories of hurricanes; she was also familiar with Mayan mythologies and the Mayan and Taino folktales about hurricanes, which she presented in storytelling workshops arranged by the Toronto Gladstone Library in Bloordale Village; I had the opportunity to enjoy the unique experience of being present there.

According to the legends of the Mayan civilization in Latin America and the Taino civilization in the Caribbean, the God and Goddess play a key role in the formation of the universe. The Mayan and Taino goddess Atabei created the earliest form of dark Earth along with Sky. Goddess Atabei had two sons, Yucaju and Guacar. Yucaju created the Sun and the Moon in the sky to provide light to the dark planet called Earth created by his parents, with light during both day and night hours or time. He further created living entities such as plants and animals to change the Earth from a dark to a living planet.

Furthermore, Yacaju created a transitional form between God and humans called Locuo to maintain an equilibrium between the Creator and its Creations. But all these positive creations of Yucaju made his brother

Guacar jealous, and it transformed Guacar into the evil god, or the god of destruction, also recognized as Huracan or Zuracan. Huracan engaged in the merciless destruction of Earth and all living entities by creating severe cyclonic storms and floods. Legends describe Huracan as a one-legged Mayan and Taino deity with a head and serpentlike body ending with a tail.

According to Mayan mythology, Huracan is one of the 13 gods central to the creation of humans. His name is derived from the word "Jura," which means one-legged. He has a head and torso, and his legs end up as one. The word hurricane is derived from Huracan.

According to Mayan legend, Huracan was engaged in three cycles of creation and destruction of Earth and its living entities.

In the first cycle, everything was made from clay, but all of his creations melted in water and ultimately perished.

In the second cycle, Huracan created humans out of wood, which then turned into manikins (small-sized humans) and angered the Mayan gods. An angry Huracan decided to destroy these manikins. To do that, Huracan unleashed a huge flood to submerge the Earth and all his creations along with it. After the destruction of his second creation, he invoked the Earth until it rose above the flood water.

Genevieve's childhood stories got me wandering in an amazing world where the dust of rocky terrain is blown away by the trampling hoofs of the running bison, where the storms of dust mixed with the last rays of the dusky sun, the fireflies danced while the wolves howled, hiding behind dark bushes of water hickory, and American elm in the moonlit night created images of a troupe of ballerinas in silver-clad black gowns. Every story of hers mesmerized me in such a way that I felt lost in the labyrinth of magic realism.

However, when Ronen came up with a marriage proposal to Genevieve to become his better half, it was not an easy decision for either of them because of their different ethnoreligious backgrounds; but it worked. I need to point out that beyond the fact that I was their next-door neighbor and close friend as well, as a person, Ronen was one of the most flexible, convincing, and gentle persons you could ever meet, one who hardly ever disagreed with Genevieve on anything she wanted to do or any decision she made. They were truly made for each other.

It was not surprising to me when Genevieve asked me for my suggestions on a nickname for Eleanor, and I was so emotionally touched by her birth that I shared with them my emotional thought, that I could not think of anything else except the picture of a dancing butterfly. Genevieve picked that one immediately and named Eleanor "Kamama," the Cherokee Indian equivalent of the same. I always called Eleanor Kamama, and as far as I remember, when Kamama started to speak, she called her mother "Ama" and Ronen "Pa," and I was "Subi."

Perhaps in my subconscious mind, Kamama had become a part of my life, but that kind of emotional attachment is not good sometimes. So when Ronen and his family relocated to Oakville, it was difficult for me to accept that every day before going to work, I would have no chance to wish Kamama a good morning.

When I used to come back from work, little Kamama would come to me and say "Subi" and stretch her palm in front of me to get her strawberry-flavored candies. So after their relocation, I went through a state of seclusion. It took me 2 years to come out of that overwhelming emotional state and make the trip to visit Ronen and Genevieve and our butterfly Kamama as I have already mentioned.

From her childhood, Kamama has been raised with a unique cultural blend of two distinct ancestral lineages, and I often volunteered to contribute to her upbringing from my end, hoping that it would help her become a good person with a broad heart.

After 2 to 3 h of primary treatment, I felt much better, though I felt a little numbness across my hands and legs and couldn't stretch them. The doctors assured me the numbness would go away after 5 to 6 h; it happens because of the neurotoxicity of the spider's venom. So I requested that if it were possible, I could be allowed to go home (here I meant my friend's home), but the doctor and health workers affirmed that it would not be possible, as they wanted to observe me overnight to make sure that my medical condition was stable, and then they would allow me to go home in the morning.

As I did not have a choice but to comply with the advice of medical experts, I requested that Ronen take Eleanor and Genevieve home and take rest; they could pick me up in the morning. I failed to convince Eleanor to go back home along with her parents; she was very stubborn and insisted on staying with me at the hospital till the morning.

So Ronen and Genevieve did not have any choice but to stay over in the waiting lounge till morning. I felt really bad that they had to do that, even though I wasn't directly responsible for such a situation. Though I felt a little weak, I started feeling better after 2 to 3 h; however, the numbness in the lower half of my body still prevailed. When Eleanor came to know that I was getting better, she insisted on coming to see me and to talk. Though kids are only allowed to move around the hospital under the supervision of their patients, by virtue of Ronen and Genevieve's professional identities, the officials allowed her to talk to me alone.

For that little exception, I consider myself lucky, as our Kamama shared an American aboriginal folktale from her collection that transported me to a world of magical realism and helped me stop focusing on the dreadful creature that had stung me. Kamama shared a classic Cherokee Indian folktale that mesmerized me; in that half-asleep state, I had the opportunity to explore this beautiful mythological world through the eyes of a child with American Indian cultural inheritance (Keams and Bernadin, 1997; Tambornino, 2010).

Kamama's Cherokee Indian folktale was about the maiden effort of the animals in the prehistoric time to bring the first fire to get rid of the darkness, which started when the Council of Animals held a meeting and passed a resolution that those who were competent enough to fly and swim should go to steal the fire from the distant island. The Council sent out the white-feathered raven to go and bring back the fire from the stump of the sycamore tree of the distant island. So after a long flight, the tired raven perched on top of the sycamore tree but failed to get close to the stump of the tree because of the scorching heat. The engulfing dark smoke rendered the fluffy white feathers of the raven black. So the raven came home with permanently dark feathers after failing the task.

The Council then decided to assign the job to the screech owl. To accomplish the mission, he flew in the direction of the glowing island, where the big sycamore tree burnt on. When the screech owl approached the hollow stump of the tree, a sudden ball of fire surged out from the base of the trunk and almost blinded the screech owl. The blood vessels in its eyeballs ruptured, making them permanently red.

Next, the hooting owl and horned owl volunteered, but both of them failed because the scorching heat of the fire and thick gray smoke impaired their vision too, so they weren't able to come out in bright sunlight. They were rendered nocturnal (becoming physically active at nighttime), and they now had a permanent white circular mark along their eyelids. The Council of Animals reviewed their early decisions to analyze the reasons for failure and finally chose a small snake and a large black snake to attempt to steal the fire. They swam across the lake to the island and managed to get into the hollow stump through a tiny hole at the base of the stump. But as soon as the snake approached the fire, it scorched the snakes' skins, and they struggled to get back home.

When most of the male members failed in their mission to get the fire, everyone was upset because they feared they would not get the warmth of fire ever and would probably never see light. The Council members decided to make one last, different effort and decided to choose a female instead of a male. Initially, it seemed an impossible task for a female entity when all the male members had already failed in the mission. Everyone looked at one another, but nobody was ready to be a part of the mission. When the Council members were about to abort the mission, a bold, beautiful tiny water spider with smooth black hair and red stripes stepped forward.

Everyone was concerned about her size, but the water spider was very much determined to do it, as her most advantageous feature was to spin a web by herself, a characteristic not found in the other animals. The water spider was able to spin a small bowl that would not burn or melt down when exposed to the heat of the fire. She tied that tiny fireproof container to her back, crossed the lake, and reached the stump of the sycamore tree. Ultimately, the water spider managed to grab two tiny bits of burning charcoal and placed them in the container.

For a tiny water spider, it was not a very easy task to traverse the long way she had come with the weight on its back. As she struggled to crawl up the mountaintop to finish her task, she threw away the large piece of burning charcoal into the sky, and it stayed there as the biggest star in the sky, the Sun.

The Sun then started giving light and warmth to all the animals and people in the daytime. She then threw away the tiny chunk of burning charcoal, which became the Moon and started giving light at night. Finally, she threw away the remaining tiny fragments and charcoal dust into the sky to clean the bowl, and they became suspended in the sky as stars and starlets.

From the beginning to the end of this pleasant and dramatic narration by Kamama, I was so amazed at the well-spun web of magical realism that I almost forgot about my physical ailments. I saw the sun rise in the hospital bed along with Kamama and felt quite thrilled to be touched by the rays of light. I realized once again how beautiful our world was, and my initial hard feelings about the black widow spider had gone away. After being released from the hospital, I returned to Ronen and Genevieve's place.

We were convinced that we could not start building the water garden immediately after spending a sleepless night, so we took naps and started the project in the afternoon. It took a little longer to finish the project successfully. At that time, I learnt another hard lesson; even though I tried hard to stay away from Kamama and maintained some distance, I failed immensely in detaching myself from her.

Sometimes, dreadful experiences create opportunities to judge and understand our stereotypic ideas of good and bad things and push us to look at things from a new perspective. Some endeavors shake up stereotypic perceptions and show how biased some beliefs are; apparently, some clarifications are very simple, but the experience leaves an immense impact in our minds. Our Kamama's narration of the Cherokee Indian folktale about the water spider further intrigued my mind to explore the origin and evolution of Akkabish and understand this creature's journey on this planet Earth.

References

Avila, L.A., Cangialosi, J., 2011. Tropical Cyclone Report Hurricane Irene (AL092011) 21–28 August, 2011. United States National Hurricane Centre.

Emanuel, K., 2005. Increasing destructiveness of tropical cyclones over the past 30 years. Nature 436. https://doi.org/10.1038/nature03906.

Keams, G., Bernardin, J., 1997. Grandmother Spider Brings the Sun: A Cherokee Story. Rising Moon, p. 32.

Oblack, R., 2020. 9 Best Hurricane Tracking Charts. ThoughtCo. thoughtco.com/find-best-hurricane-tracking-charts-3443940.

Rieti, J., 2009. 'Just a Silly Spider Bite' Leaves Senior Paralyzed. https://www.thestar.com/news/gta/2009/08/13/just_a_silly_spider_ bite_leaves_senior_pa ralyzed.html.

Saunder, M., Lea, A., 2019. Pre-Season Forecast for North Atlantic Hurricane Activity in 2019 (PDF). Tropical Storm Risk, London, United Kingdom. Retrieved June 3, 2019.

Tamborino, P.D., 2010. Maggie's Story: Teaching's of a Cherokee Healer. Mammoth, p. 136.

CHAPTER TWO

An exordium to the cultural delineation of spiders

Abstract

Since ancient times, the spider has occupied center stage in diverse customs, sociocultural rituals, spiritual practices, age-old beliefs, traditional entertainments, depictions of literary and artistic intricacies, etc. The spider has played a key role in the life-changing processes of historical characters such as King David, Lord Buddha, Prophet Mohammed, Japanese shogunate Yoritomo, and Robert the Bruce, the great warrior of Scotland, and thereby this mythological catalyst has orchestrated the course of history. From African Anansi folktales to Australian indigenous paintings, spiders have been depicted as sacred spiritual bodies to the aboriginal communities. North Americans, specifically the native Indian communities, have regarded spiders as a mythological legend, a symbolic sociocultural emblem. In Europe, however, this frightful creature has been depicted as part of the evil spirit, the typical socioeconomic equilibrium since the dawn of civilization.

Keywords: Mythological legend; Sociocultural emblem; Spider; Spiritual entities.

Fiction is like a spider's web, attached ever so slightly perhaps, but still attached to life at all four corners. Often the attachment is scarcely perceptible.

Virginia Woolf

In the course of historical depictions of famous characters, including one of the most powerful rulers in Middle East Asia—King David of Israel; the founder of Islam—Prophet Mohammed; the founder and first shogun of the Kamakura shogunate (the most powerful military leader in ancient Japan)—the great conqueror Yoritomo; and the great king of Scotland—Robert the Bruce, historians have relied on intensive research of age-old scriptures, archaeological artifacts, unique historic evidence, etc. Apparently, historians could not find any common issue to narrate the glorified journeys of success of these historic characters in terms of space and time, and their distinct pattern of achievements would not have any analogical resemblance (Scott, 1928; Cloudsley-Thompson,1968; Zerah, 2000; Zain, 2019). Yet there was a common factor that steered the course of the journey of history and its further advancement by the actions of the aforementioned luminaries on

Spider Evolution
ISBN: 978-0-323-90041-6
https://doi.org/10.1016/B978-0-323-90041-6.00007-3

this Earth—nothing but a most humble creature, the spider. Its wisdom is believed to be greater than that of the whole world (in accordance with the traditional belief of the people of the Akan ethnic group in West Africa) (Rattray, 1930).

Since prehistoric ages, the spider has been one of those omnipresent entities that has marked its presence in mythological folktales, in the oral fables that are famous in sociocultural rituals, and in traditional faiths based on spiritual practices. One of the old mythological stories related to the role of a spider in human life was rolled out by the Torah around 3060 years ago when the army of King Saul of Israel was pursuing King Saul's enemy David (who later became King David, one of the most powerful kings in Israel), who took shelter inside a cave at Jerusalem (Zerah, 2000). A spider immediately spun a silk web over the opening of the cave, which was sufficient to misguide the soldiers of King Saul into believing that the cave was abandoned, and thus they did not bother to search that cave in their search for David. So the timely action of a spider, which has been considered as acted upon by the sublime wishes of God, helped save the life of David, who would later become known to the world as the most powerful king of Israel.

The next miraculous action of a spider happened to a person around 1400 years ago who has been recognized as the prophet Mohammed, the founder of Islam (Zain, 2019). There was a popular story woven around CE 620 when Prophet Mohammad along with his main disciple, Abu Bakr (leader of one of the leading tribes in Arabia at that time), and a number of followers undertook his migration or Hegira from Mecca to Yathrib (later recognized as Medina). He was being chased by Quraysh soldiers and took shelter in the Cave of Thawr to protect himself from his enemies. At that time, a wood pigeon had nested on an Acacia tree, and a spider built a web between the tree and the cave entrance that blocked the view of entrance in accordance with the instructions of Allah. Prophet Mohammed's enemies could not locate him in his hideout, and later he preached the message of Islam all over the world.

An age-old mythological story is set in Japan around 800 years ago when it was reigned by military clans, specifically by the great samurai leader Minamoto Yoritomo of the Minamoto clan and his loyal samurai army (Cloudsley-Thompson, 1968). The popular Japanese story entailed the Battle of Genpei when Yoritomo was chased by rival members of the Taira clan and took shelter in the hollow tree trunk of a robust tree and prayed to God to save him. The messenger of God instructed a spider to spin a web at the opening of the hollow tree, and the spider did. The web confused the enemies of the Taira clan, who abandoned the search operation for Yoritomo, as they had no clue that Yoritomo could hide inside the hollow stump of a tree that looked like it had been without human touch for a long time. The escape gave Yoritomo a new lease on life, and he became a powerful military leader or shogun in the reign of the Minamoto clan in Japan. Finding its seeds in ancient days of Japan, one of the Japanese traditional stories described "Jorogumo" as a prostitute spider who took the shapes of beautiful women and seduced the Japanese samurai in the name of marriage. But at the same time, "Jorogumo" or "Joro-Spider" has also been regarded as a goddess by travelers traveling through the Joren Falls location of the Izu city peninsula, as she saved their lives from further drowning in the falls. According to another popular traditional story, Minamoto no Raiko, the Japanese legendry, had an opportunity to meet "Tsuchigumo," the earth spider, who had the magical power of metamorphosis to take the shape of a male or female figure. It has further been told in that traditional folktale that Minamoto was on a mission related to the search for the mythical "giant skull," and was illusioned by "Tsuchigumo," who confused Minamoto in the shape of a young male, and whenever Minamoto came out of that elusive trick of the earth spider, found himself entrapped in the spider's web.

Around 750 years ago, Robert the Bruce, the most famous Scottish warrior and who fought the first war of Scottish independence against King Edward I of England, was on the verge of defeat and lying in a barn, physically and mentally exhausted by huge casualties among his soldiers. Somehow, he happened to look at the ceiling of the barn and noticed a spider trying to fix its web, which was attached to a ceiling beam. What he noticed was that every time it tried to fix the web, the spider failed. Robert kept watching it inquisitively, noting that although the spider failed every time, it did not give up, making effort after effort; finally, in its seventh attempt, it tasted success (Scott, 1928). That intermittent endeavor of the little creature boosted Robert's morale, and he fought back with greater vigor and won the first historical battle of independence of Scotland fought against England and also against the well-trained army of Edward I in 1319. All these mythological and ancient historical stories prove that spiders have played an important role in the sociopolitical history of human civilization since its dawn on this Earth.

According to a popular Japanese folktale, titled *Spider's Thread*, by Ryunosuke Akutagawa, Lord Buddha was walking around the end of the Lotus Pond in paradise when he noticed a man with an evil-spirited bandit, named Kandata, was struggling with other sinners to come up from the bottom of hell, which was a blood-filled pond (Akutagawa, 2012). Although Kandata had performed mostly evil deeds throughout his life, he had once done a single good deed by sparing one tiny creature without trampling it when he was hanging out in a deep forest. Remembering that good deed by Kandata, Lord Buddha became compassionate and decided to give him a chance to be out of hell. At that very moment, Buddha noticed that a spider of paradise had weaved a web made up of the silver threads on the surface of a lotus leaf. Lord Buddha took the silver thread and descended straight down to the bottom of hell to make sure that Kandata would be able to hold that silver thread securely and come out of hell to reach heaven. Kandata held the thread and struggled to come out of hell toward heaven. Eventually, Kandata had been partially successful by coming halfway up toward heaven, and then he looked back and observed that a large number of humans were following him in his struggle to come out of hell, which made him scared that the silver thread might not be able to bear the burden and would break. So Kandata shouted to the sinners below him and claimed that the silver thread belonged to him and told them to leave the thread. Immediately after that behavior, the thread snapped off from the exact portion held tightly by Kandata, and he dropped off to hell again. This Japanese folktale revealed that Lord Buddha had shown his compassion even to an evil-spirited bandit to redeem his soul by coming out of hell, but the selfishness of some people, like Kandata, would always remain because of their selfish attitudes, and thus such sinners were dragged down to hell again.

The spider played an important role in ancient, traditional Chinese sociocultural life. In China, the spider has been considered the symbol of happiness and has been called "Xi Zi," which made Chinese people consider it a good omen. They have believed in the supernatural power of the spider since the Han dynasty. Specifically, Chinese people believe that the spider always falls out of its nest on someone to make him wealthy. As a result, a spider hanging from a web became a famous form of symbolic painting/artifact considered to bring fortune (Vision Times Chinese, 2014). It might have been considered simple traditional practice in the old days, but Chinese women put two chopsticks inside a container on the seventh day of the seventh month in the nighttime to find out whether a woman was skillful enough in stitching and weaving. Clothing is based on the findings of whether a piece of cobweb there adhered to the chopsticks the next morning. In 1592, Wu Cheng'en, a famous novelist, depicted spiders as demon damsels in his classic novel, *Journey to the West*, where he depicted a Chinese monk, Tang Sanzang, who traveled from China to India along with his three disciples, monkey (Sun Wukong), piggy, and sandy (a lake demon) (Lovell, 2013). On their way to India from China, they came across several demons who wanted to either devour Tang or marry him, as Tang Sanzang was believed to be a former god who was reincarnated as a human with special powers like an immortal entity. So devouring Tang would help any demon enjoy immortality, and Tang was handsome as well, so even the female demons wanted to marry him to get in touch with his special powers. In the middle of their journey, Tang Sanzang and his disciples met seven beautiful sisters in a silk cave and asked for food without knowing that the women were spider demons. They had tricked Tang as beautiful damsels to try to imprison him to either eat him up or marry him. Once Tang realized the bad intentions of the spider demons, he tried to escape but was overpowered by the seven sisters, who entangled Tang in their web. Sun Wukong (monkey), the valiant disciple of Tang, fought a hard fight and rescued Tang from the clutches of the spider demons, but the seven sisters tried to recapture Tang. So Sun Wukong killed them all and moved forward with the monk Tang Sanzang on the way to India. So in the Chinese folktale, the spider has been depicted as a demon, whereas it has mostly been depicted as a goddess in other traditional sociocultural depictions.

Most of us are not familiar with the name "Anansi," which is the name of a tricky spider. It was originally mentioned in stories from the Ashanti and Akan people of Ghana, from where they became the stories of West Africa (Rattray, 1930). According to the colloquial West African folklore, Anansi (which supposedly originated from the Kwaku Ananse in West Africa and was changed to the anglicized name of Aunt Nancy in the Caribbean), a cunning trickster, took the shape of a spider able to hold all stories and folktales and act as a messenger between God and the people. At the time of the slave trade, all these folktales drifted out of the African continent and came across to the New World as the oral cultural heritage infused in the collection of a treasure trove of folktales in North America to enrich the existing collection of folktales. Anansi's folktales inspired the African slaves in the Caribbean to organize resistance against the slave traders. According to one Anansi folktale, Anansi, the messenger between God and people, was once ordered by the Sky God to spin a kind of fabric web from which the people could be made. The Sky God also created the day, night, rain, and wind for the people as part of its creation of the world. According to another Anansi folktale, Anansi had gathered all the wisdom in a big container, but it was too big for him to carry. Anansi's son recommended that he carry the big pot on his back, but Anansi did not hold it properly, so a big chunk of wisdom trickled down out of the container and spread out randomly all over the world so that people would have a chance to get in touch with it and become wiser.

In Australian indigenous paintings like bark paintings and rock paintings, the aboriginal communities of the Northern Territory of Australia portrayed the spider as an integral sacred, spiritual body in their sociocultural and religious life (Berndt et al., 1953). According to the cultural heritage of Australian aboriginal communities of the Rembarrnga, or Kyne, group, spiders are considered one of the most sacred totems,[1] known as the "Burungku clan totem" in their "Central Arnhem Land" (Evans, 2012). These aboriginal communities consider spiders to have integrated their fundamental regional and sociocultural ceremonies in their webs. The spider totems represent the symbol of solidarity between the Burungku clan with other neighboring clans, who also regarded the spider totems in performing their sociocultural and religious functions and rituals (Reid, 2005).

The sketches of spider figures in the paintings and body tattoos of the aboriginal communities in the Oceanian islands of Vanuatu, specifically the Banks Islands in its northern part (Bonnemaison et al., 1996; Evans, 2010; Frost, 2017). In the southern part of Vanuatu, particularly Malakula Island, the spider web has been recognized as a form of sacred artwork and part of the smothering hoods or "cap of death." On many occasions, it has been described that in ancient times, various demonic practices were involved when widows were suffocated to death at the funerals of their husbands. The aboriginal people have also considered spider spirits as playing an important role in representing the immediate and transitional phases of life after death. The aboriginal community members of the Malakula Island of Vanuatu have reportedly walked through the bushes in the morning and collected spider webs by pulling them down from the crowns of trees with long poles, which then have been utilized as waterproof, thick, special clothing materials and reportedly used in socioreligious rituals. The aboriginal communities of the small Micronesian island state the Republic of Nauru believed in the traditional cosmology and that "Areop-Enap," the old spider, was the creator of the universe (Bartlett, 2009). The aboriginal communities of the Tungaru archipelago (Gilbert Islands) of the Republic

[1] Totem: It is an animal figure or natural body, which has been adopted as an emblem, considered to be of substantial spiritual significance to the aboriginal communities of any particular country concerned.

of Kiribati (a small Micronesian island state) between the Oceanian islands and Hawaii in the Pacific Ocean believed that "Nareau" was the creator of the universe (Brewis, 2012).

According to Sumerian myth, in the ancient civilization of Mesopotamia, inhabitants known as Sumerians performed the age-old religious practices called Sumerian religions, and according to that ancient Sumerian traditional belief, "Uttu" was one of the important goddesses involved in spinning a web with an analogical resemblance to the spinning of the fate of humankind (Black and Green, 1992). Goddess "Uttu" was described as the daughter of the Sumerian water god "Enki" and his wife "Ninmah." Once, Ninmah warned Uttu about her frightful premonition that Uttu would be a victim of the sexual misconduct of Enki by means of seducing her, which made Uttu scared about her normal course of living, and she confined herself in her own-spun web (Jacobsen, 1987). However, Enki was successful in seducing Uttu by luring her with food and beverages and promised her marriage (Jacobsen, 1987). Enki gave Uttu beer until she became inebriated and then committed forceful sexual misconduct against her will, and her screaming for help made "Ninmah" rush to her to rescue her and prompted her to cleanse her reproductive part free from the semen planted by Enki, and when it touched the ground, eight unique plants grew from it (Jacobsen, 1987).

In Egyptian mythology, the spider has a spiritual inheritance from the traditional goddess "Neith," who was engaged in weaving or spinning the web of the destiny of humankind, and it has a symbolic resemblance to the Babylonian god "Ishtar," the Celtic goddess "Arianrhod," the Greek goddess "Athena," the Roman goddess "Minerva," etc. According to Celtic legendry, spinning, as well as weaving, of the web by the spider has a mystic connection to the act of God, which has yet to be fully explored. The Celtic myths further enthralled the mystic identity of the goddess "Arianrhod," who has a strong inherent spiritual connection to a spider and is involved in weaving the fate of humankind (Shaw, 2013).

According to Greek mythology, "Athena" is the goddess of wisdom, crafts, and warfare and considered the equivalent of the Roman goddess "Minerva." She faced a stiff challenge from a woman known as "Arachne" who claimed she could work better and faster than Athena (Evans and Millard, 1985). So Athena accepted Arachne's challenge and participated in a weaving contest, where Arachne's work was appreciated for her faster spinning abilities and the better finishing qualities of her spun tapestry than those of Athena. The main theme of the tapestry woven by Arachne was a depiction of "infidelity among Gods of Olympus," and Athena became infuriated by the mortal audacity of Arachne (Mills, 2003). That made the goddess Athena jealous of her rival Arachne and led to her destruction of the beautiful artifact Arachne had made. Arachne became so upset that out of grief, she decided to commit suicide by hanging herself. This made the goddess Athena change her mind and diffuse her anger. She used a miracle herb called aconite to bring Arachne back to life, but simultaneously she transformed the hanging rope into a cobweb and metamorphosed Arachne into a spider engaged in weaving the web from that point forward, always hanging on its cobweb (Bulfinch, 2014). Since then, the spider has been known as Arachnid to all of us. This Greek mythological story was narrated by Virgil, the great Roman poet in the Augustan regime, in 29 BCE and was further elucidated in the poem of the great poet Ovid, who mentioned the Greek mythological depiction of Arachne in his poem titled "Metamorphoses" between CE 2 and CE 8 (Ovid (The Metamorphoses of Ovid, translated by Allen Mandelbaum), 1993).

Though the sociocultural interpretation of the spider symbol varies from diverse, aboriginal groups of Native American Indians spread throughout North America and further extended to Latin America, it is considered a sacred symbolic form of mysticism in their aboriginal foundation. The spider is one of the illuminating cultural foundations of the Native American cultural treasure trove in North America and has been recognized as the common and sacred sociocultural emblem among the different Native American Indian aboriginal groups throughout North America. Native American mythology has depicted the spider as a legend and the originator of the constellation Ursa Major, when the spider transformed seven people to seven stars, and they were provided the well-spun web as a ladder to traverse to paradise (Garai, 1974).

The native American aboriginal group known as Lakota of Dakota, USA, considered the spider trickster "Iktomi" their sociocultural hero, who was the son of "Inyan" (the rock), one of the powerful male gods, and it was believed by the Lakota American Indian tribe that Iktomi used to have a pearl of wisdom (Ksa) that had been taken away because of his troublemaking attitudes (Lloyd, 2019). According to Lakotan mythological folklore, Iktomi had the magical power to take any shape. Although he primarily took the shape of a spider, he later took the shape of a human painted in bright red, yellow, and white with dark black rings around the eyes. This depiction has been so common that Iktomi is recognized as Spider-Man in the Lakota American Indian aboriginal communities, as this trickster spirit has the potential to change immortal entities (like gods) and mortal entities (like a humans) by means of applying "potion" (the magical liquid drug having metamorphosing potential to establish its control over mortal or immortal entities) (Zitkala-Sa,1985). According to the Hopi Tribe of Arizona, USA, "Spider Woman" is the oldest goddess on Earth, and "Tawa" is the sun god, and they have existed since the formation of Earth in those days when there was no life on it (Alexander, 1996). That intrigued the goddess "Spider Woman," and she became the partner of "Tawa" to take part in the process of creating human life on Earth (Wigington, 2018). It has further been contended that their composite actions were able to form the first man and woman on Earth, where Tawa planned the distinct structure and shape of the first man and woman, while the Spider Woman physically created it from mud out of a clay mold (Wigington, 2018). The famous Cherokee folktale regarding the formation of the illuminating Earth along with the formation of other celestial bodies like the sun, moon, and stars, and the unique roles played by water spider or Grandma Spider, was discussed earlier (Wilcox, 2007).

The nature worshipping prehistoric people in the ancient Moche civilization in Peru between CE 100–700 traditionally worshipped the animal gods and depicted spiders as a holy spirit (Berrin, 1997). At the same time, along the southern coasts of Peru, archaeological explorations of geoglyphs (the large motif or landscape design of rocks on the ground) of spiders in the Nazca desert and Ica Valley have yet to elucidate, from archaeological and anthropological perspectives, the geoglyphs popularly known as the "Nazca Lines" (Jarus, 2012; Golomb, 2019). In the archaeological exploration of the Lambayeque region of the North Pacific coasts of Peru, the temple (named "Collud") of the spider god has been explored, which was supposedly established from 1500–500 BCE, during the Cupisnique civilization (Talpalariu, 2008). The spider deity was reported to be associated with power, war, etc., as the ancient pictorial evidence elucidated that the spider deity holds hunting nets full of decapitated human heads (Orozco, 2008).

In traditional neopagan beliefs, the spider is regarded as the "Goddess of Creation," with enormous power of wisdom (EarthSpirit, 2017). So in its traditional practices, neopaganism incorporates spider webs as a traditional method of meditation to get imbued with the inward streaming of this energy stored in the "Goddess of Creation." According to the old English traditional belief, to find a spider on the clothing means that fortune will come to the person the clothing belongs to in terms of suddenly gaining substantial wealth.

Every coin has two sides; likewise, not every mythological story related to the spider can narrate positive messages. As mentioned in the *Encyclopedia of Witches, Witchcraft and Wicca* related to folk magic, successfully eating a black spider between two slices of buttered bread will enhance a witch's black power (Guiley, 2008). Meanwhile, some practices of aboriginal rituals involved entrapping the black spider in a silk pouch and using it as the bead of a necklace to help people avoid a number of ailments caused by bad spirits.

According to Polish legend, folk hero Pan Twardowski was a nobleman who lived in the city of Krakow during the 16th century who reportedly sold his soul to the devil (having a close resemblance to Faust of Germany) to achieve black magic and supernatural power (Kepa, 2019). Eventually, Pan Twardowski became a courtier of the king Sigismund Augustus, and he pleased the king by exhibiting his special power by mesmerizing the courtyard with his magic tricks and astrological foretelling. Pan Twardowski reportedly brought back the soul of the dead queen Barbara Radziwill using a magic mirror to soothe the grieved king. But he had been tricked by the devil, and his soul got stuck in the magic mirror and dragged by the devil destined to hell. On his journey to hell, he began to pray to Mother Mary to help him, and he was dumped on half of the moon and became stuck there forever. It has been construed further that once Twardowski had transformed his assistant into a spider, who became his companion because he lived on the moon, and he requested that the spider act as a messenger to pass the messages of the world of a living being to him in the world of the moon (Widacka, 2009). It was an interesting East European folktale that depicts the spider as a messenger between the Earth and moon.

The earliest narration of science fiction, written in the 2nd century CE, is one of the finest presentations of mythical events like aliens, their participation in interplanetary warfare, etc. These stories are nicely portrayed in *A True Story* by Lucian of Samosata, a Greco—Syrian author of an indigenous ethnic group of Assyrian (Middle-East) descent (Georgiadou and Larmour, 1998). Lucian's novel was a unique depiction of the metaphoric battle between the King of Sun and King of Moon regarding disputes that could arise in the process of colonization of stars, where Sun has won over Moon followed by a settlement through a peace agreement between the warring parties. In his unique novel, Lucian describes the mythical portrayal of the armies participating in this mythical war and describes their hybrid-looking form with a fine blend of imagination and unique ideas. Also, a giant spider was center stage in his unique endeavor to begin the creation of science fiction in the West in its earliest form during those days of the 2nd century CE (Casson, 1962).

In 1842, Jeremias Gotthelf, the Swiss novelist, nicely presented the conflict of good and evil in his popular Christian-humanist allegorical novel *The Black Spider*, in which a peasant's wife, "Christine," tries to help out fellow peasants living around the foothills of the Alps in Switzerland and Bavaria (Gotthelf (The Black Spider, Translated by Susan Bernofsky), 2013). However, those peasants were found to be marginalized in the face of tyranny and socioeconomic oppression of the feudal lord in the 17th century, Hans von Stoffeln, and Christine surrendered to the devil and was transformed into a demonic spider who killed not only a section of poor farmers and the cattle, but also Hans von Stoffeln, to end the hunger and reign of terror and social oppression (Gallagher, 2008). Eventually, the demonic spider was killed, and the will of God prevailed to bring peace to the lives of the villagers. Jeremias Gotthelf's novel has portrayed the demonic spider as part of the evil spirit, yet this creature has been portrayed as the key factor in maintaining socioeconomic equilibrium since the beginning of civilization.

In 2001, J. K. Rowling's *Fantastic Beasts and Where to Find Them* dealt with unique depictions of a number of magical creatures. In the universe of Harry Potter, where the magical, giant-sized spiders were recognized as "Acromantula," the plots of the novel portrayed the magical world against the backdrop of Harry Potter's unique experiences during his studies/internship in his second year at Hogwarts School of Witchcraft and Wizardry (Rowling, 2009).

The great American comic book writer Stan Lee, alias Stanley Martin Lieber, gave birth to a comic hero, "Spider-Man," which was published for the first time during the "silver age of comic books" in August 1962. Stan Lee had depicted the central figure of this comic, Peter Parker, as an orphan who had been raised by his Aunt May and Uncle Ben in New York City and a kid who struggled in childhood, a symbolic figure brought up in the struggling lower-income group of people living around New York City (like the kids growing up in Queens or the Bronx). The transformation of Peter Parker was from a freelance photographer cum scientist to a "Spider-Man" with superhuman potential imbued by an accidental bite from a genetically engineered/radioactive spider. The spider infused the potion in Peter Parker, rendering him the acquired special magical power to scale high-rise buildings. It should be mentioned that this comic was so popular in North America and other parts of the world that cinematographers made a television film based on it that has been produced and telecast by the CBS network, followed by the production of its full cinematic version in three series in Hollywood as *Spider-Man* in 2002, *Spider-Man 2* in 2004, and *Spider-Man 3* in 2007 (Aguilar, 2019). The take-home message of this most famous comic was that Spider-Man started using his special powers to fight all evil forces, which again played a cardinal role in depicting the positive role of an apparently weird creature like a spider in maintaining the sociocultural equilibrium of society.

So whether it's an ancient folktale, a novel of ancient times, a recent science fiction book, or a contemporary comic series, the existence of the spider as a prime character in a transitional phase from sociocultural changes through time in different parts of the world has always occupied center stage in understanding who we are in our ambiance and helping us reach the doorstep of sociocultural evolution, and it has maintained its dynamic identity to traverse through changes of time and space. In the age of the cyber-revolution, the spider's web has become a metaphoric but integral part of our social life, as our virtual connections to the rest of the world (whether near or far) are literally impossible without the interconnectivity of the spider's web, the World Wide Web (WWW), popularly known as the Internet. The groundbreaking invention of the World Wide Web by the British scientist Tim Berners-Lee in 1989 not only ushered in the regime of cyberrevolution but also entangled 4834 million people in the world out of around 7800 million total, indicating that around 62% of the global population is entangled in an invisible megaweb in this age of exploring paradigms of information technology when human civilization has traversed through the evolutionary trail of contemporary digital age paved by the Fourth Industrial Revolution (Thomson, 2015; Schwab, 2016; Internet World Stats, 2020).

That is why being a part of the contemporary society, which came a long way from the spatiotemporal trail of Anthropocene and the era of abrupt climate changes mainly due to anthropogenic factors (Ellis, 2018), contemporary sociocultural and scientific progress has intrigued us to further know and understand that "inconsequent" creature with immense sociocultural importance that is apparently weird looking, as it has always created a buzz in various times and various parts of the world by raising the confiscatory issues between good and evil, and keeping the issues afloat, it has intrigued us to be inquisitive and draws our attention to explore its origin and evolutionary importance on Earth.

References

Aguilar, M., 2019. 'Spider-Man: Far from Home' Reveals Peter Parker's Birthday and 'Amazing Fantasy' Easter Egg. https://comicbook.com/marvel/news/spider-man-far-from-home-birthday-amazing-fantasy-easter-egg/.

Alexander, T., 1996. The fourth world of American philosophy: the philosophical significance of native American culture. Trans. Charles S. Peirce Soc. 32 (3), 375–402 (Summer 1996).

Bartlett, S., 2009. The Mythology Bible: The Definitive Guide to Legendary Tales. Sterling, New York, p. 176.

Berndt, R.M., 1953. Australian Aboriginal Art. Ure Smith, Sydney, NSW p. xiii+117.

Berrin, K., 1997. In: Berrin, K. (Ed.), The Spirit of Ancient Peru: Treasures from the Museo Arqueológico Rafael Larco Herrera. Thames and Hudson, Inc, New York.

Black, J., Green, A., 1992. Gods, Demons and Symbols of Ancient Mesopotamia: An Illustrated Dictionary. The British Museum Press, London, England, p. 182.

Bonnemaison, J., Kaufmann, C., Huffman, K., Tryon, D. (Eds.), 1996. Arts of Vanuatu. Crawford House, Bathurst, NSW.

Brewis, A., Crismon, S., 2012. "Kiribati": Countries and Their Culture. Advameg, Inc. https://www.everyculture.com/Ja-Ma/Kiribati.html.

Bulfinch, T., 2014. Bulfinch's Mythology: The Classic Introduction to Myth and Legend-Complete and Unabridged. Penguin Publishing Group, p. 730.

Casson, L. (Ed.), 1962. Selected Satires of Lucian. W.W. Norton & Co, New York City, NY.

Cloudsley-Thompson, J.L., 1968. Spiders, Scorpions, Centipedes and Mites. Pergamon.

EarthSpirit, 2017. The Boy Who Longed to Fly. http://www.earthspirit.com/boy-longed-fly#more-216.

Ellis, E.C., 2018. Anthropocene: A Very Short Introduction. Oxford University Press, New York, p. 208.

Evans, N., 2010. A tale of many tongues: documenting polyglot narrative in north Australian oral traditions. In: Baker, B., Mushin, I., Harvey, M., Gardner, R. (Eds.), Indigenous Language and Social Identity: Papers in Honour of Michael Walsh. Pacific Linguistics, Canberra, ACT, pp. 275–295.

Evans, N. Spiders in Australian indigenous art. Australian Museum. Online (Retrieved November 14, 2012) https://australian.museum/learn/animals/spiders/spiders-in-australian-indigenous-art/.

Evans, C., Millard, A., 1985. Usbourne Illustrated Guide to Greek Myths and Legends. Usbourne Publishing, p. 15.

Frost, N., 2017. These Hoods Made of Spider Webs Have Long Been Misunderstood. https://www.atlasobscura.com/articles/ooi-spiderweb.

Gallagher, D., 2008. The Spider's Bride. Willside Press, p. 248.

Garai, J., 1974. The Book of Symbols. Simon and Scuster, p. 143.

Georgiadou, A., Larmour, D.H.J., 1998. Lucian's Science Fiction Novel, «True Histories»: Interpretation and Commentary, Leiden.

Guiley, R.E., 2008. The Encyclopedia of Witches, Witchcraft and Wicca. Checkmark Books, p. 436.

Golomb, J., 2019. Nascà Lines. https://www.nationalgeographic.com/history/archaeology/nasca-lines/.

Gotthelf, J., 2013. The Black Spider. (Translation in English Made by Susan Bernofsky). NYRB Classics, p. 120.

Internet World Stats, 2020. Internet Distributions in the World-2020 Q1. https://www.internetworldstats.com/stats.htm.

Jacobsen, T., 1987. The Harps that once: Sumerian Poetry in Translation. Yale University Press, New Haven, Connecticut, p. 184.
Jarus, O., 2012. Nazca Lines: Mysterious Geoglyphs in Peru. https://www.livescience.com/22370-nazca-lines.html.
Kepa, M., 2019. Pan Twardowski: The First Pole on the Moon. https://culture.pl/en/article/pan-twardowski-the-first-pole-on-the-moon.
Lloyd, E., 2019. Iktomi – Native American Spider-Trickster Spirit Whose Stories Teach Moral Values. https://www.ancientpages.com/2019/01/28/iktomi-native-american-spider-trickster-spirit-whose-stories-teach-moral-values/.
Lovell, J., 2013. Monkeying around with the Nobel prize: Wu Chen'en's "Journey to the West". https://lareviewofbooks.org/article/monkeying-around-with-the-nobel-prize.
Mandelbaum, A., 1993. The Metamorphoses of Ovid: A New Verse Translation. Harcourt Brace & Company, p. 559.
Mills, A., 2003. Mythology: Myths, Legends & Fantasies. Global Book Publishing Pty Ltd, Australia, pp. 62–64.
Orcozo, J., 2008. Spider God' Temple Found in Peru. National Geographic News.
Rattray, R.S., 1930. Akan-ashanti Folktales; Collected and Translated by Capt. R.S. Rattray and Illustrated by Africans of the Gold Coast Colony. Oxford University Press, p. 309.
Reid, M., 2005. Mythical Star Signs. Lulu.com, Scotland, p. 128.
Rowling, J.K., 2009. Fantastic Beasts and where to Find Them: From the World of Harry Potter. Bloomsburry Press, p. 128.
Ryunosuke, A., 2012. The Spider's Thread by Ryunosuke Akutagawa. https://badianoran.wordpress.com/2012/03/22/literary-criticism-final- requirement-in-world-literature/.
Schwab, K., 2016. The Fourth Industrial Revolution: What it Means, How to Respond. https://www.weforum.org/agenda/2016/01/the- fourth-industrial-revolution-what-it-means-and-how-to-respond/.
Scott, R.M., 1982. Robert the Bruce: King of Scots. Carroll & Graf, p. 253.
Shaw, J., 2013. Arianrhod, Celtic Star Goddess. https://www.feminismandreligion.com/2013/02/23/arianrhod-celtic-star-goddess-2/.
Talpalariu, D., 2008. Peruvian Temple of the Spider God Revealed. https://news.softpedia.com/news/Peruvian-Temple-of-the-Spider-God- Revealed-96832.shtml.
Thomson, S., 2015. Is This the Start of a Fourth Industrial Revolution? https://www.weforum.org/agenda/2015/09/fourth-industrial- revolution/.
Vision Times Chinese, 2014. 5 Surprising Auspicious Chinese Symbols: Bats and Spiders Are Good Luck? www.visiontimes.com/2014/04/13/5-surprising-auspicious-chinese-symbols-bats-and-spiders-are-good-luck.html?photo=2.
Widacka, H., 2009. Legendy I Fakty O Mistrzu Twardowskim (in Polish). www.wilanow-palac.art.pl.
Wigington, P., 2018. Spider Mythology and Folklore. https://www.learnreligions.com/spider-mythology-and-folklore-2562730.
Wilcox, L.S., 2007. Grandmother spider: connecting all things. Prev. Chronic Dis. 4 (1), A01.
Zain, M., 2019. The prophet & the spider. Story Hijrah. https://aboutislam.net/family-life/moms-dads/prophet-spider-story-hijrah/.
Zerah, A., 2000. How the Children Became Stars: A Family Treasury of Stories, Prayers, and Blessings from Around the World. Sorin Books, p. 160.
Zitkala-Sa, 1985. American Indian Stories. University of Nebraska Press, Lincoln.

CHAPTER THREE

Spiders: The seventh-ranked species on planet Earth coming from the mythological abode of seven earths (or seven heavens)

Abstract

From traditional sociocultural narratives of religious practices, folklore, and fables as well as mythological depictions in ancient literature, spiders have been portrayed as fearsome, mystic creatures between heaven and Earth. Naturally, such depictions stalled scientific investigation to unfurl the biological identity of the arachnids, which evolved around 300 to 400 million years ago. The in-depth morphoanatomical and biobehavioral studies of this fearsome creature have helped scientists, particularly evolutionary biologists, recognize the spider's poor monochromatic vision (with 6–8 primitive eyes thought to recognize dark and light shade), slow movement with eight jointed legs (lack of muscles and endoskeleton), primitive central nervous system, and intracellular streaming of hemoglobin-free blood. These characteristics would not likely render it a mythological, fearsome predator to its wide range of prey (from tiny insects to large mammals).

Keywords: Arachnids; Poor vision; Primitive arthropods; Slow movement.

What I am interested in with birds, just as I am with spiders or monkeys, is what they do and why they do it.

David Attenborough

The religious practices in Mesopotamia (the traditional religious practices in Assyria, Babylonia, Sumer, etc.) from 3500 BCE to CE 400 included mythological astronomy referring to the seven levels of heaven or "seven heavens" integrated into the seven celestial entities (Mercury, Venus, the Moon, the Sun, Mars, Jupiter, and Saturn) (Horowitz, 1998; Campbell, 2016). Mythological perceptions of heaven can be found in the contemporary religious practices of Christianity, Judaism, and Islam and have also been found in slightly modified form in Hinduism and Jainism (Jansma, 2006; Ibn 'Arabi and Ibn Al'Arabi, 2008; Dalal, 2010; Corrigan et al., 2011). According to Mesopotamian cosmology, between 2000 BCE and 1001 BCE, the

Spider Evolution
ISBN: 978-0-323-90041-6
https://doi.org/10.1016/B978-0-323-90041-6.00008-5

ancient Sumerian religious practices of mysticism referred to seven heavens (an-imin-bi), considered the abode of the God and Goddess, and seven earths (ki-imin-bi), considered the abode of humans, animals, and other living creatures, where heavens have absolute sublime force/power to control the origin and evolution of living entities on Earth (Horowitz, 1998). The mythological portrayal of spiders in the ancient literature and their traditional cultural depictions in many religious practices, folklore, and fables describe them as mystic creatures between heaven and Earth. These characterizations have further added to the mystic attributions given to the spider as a scary, fearsome entity rather than a fascinating animal. Such biased, stereotypic depictions of the spider often limit the scope of naturalists, biologists, and scientific investigators to explore its biological identity. The paleontological evidence has further ascertained that this apparently fearsome creature evolved around 300 to 400 million years ago from stratified amber on planet Earth (Pennisi, 2018).

The slow-moving creepy creature, apparently scary but one of the fascinating invertebrate animals (i.e., no spine in its body), is easily recognized by its locomotory organs, which are typically characterized by eight jointed legs along with the sensory organs, six to eight eyes (ranges from sets of three to four), that aid movement and food intake. In the course of biobehavioral studies on arachnid vision, spiders have been determined to be shortsighted, which might be a reason for their slowed movement and inability to judge potential predators from a long distance. In recent studies on spider photosensitivity, scientists revealed that due to the absence of true lenses, spiders supposedly sense differences in light and shade rather than see objects with their simply structured eyes. In addition to their eyes, spiders have a simple form of "central nervous system" that helps this fascinating creature sense danger (Szalay, 2014).

Histologically, the blood of human beings, which is bright red due to the presence of the protein hemoglobin, is a distinctive feature of vertebrates, whereas the histological characteristic of spiders has a striking resemblance to other invertebrates, the color of which look a little bluish because of the presence of hemocyanin instead of hemoglobin. From a zoological perspective, the spider belongs to the phylum Arthropoda, and within the Arthropoda, it belongs to the class Arachnida, which has the greatest number of species. Although insects also belong to the arthropods, arachnids are distinct from insects in their conspicuous absence of antennae. Arachnid members do have a striking biobehavioral and physiological resemblance to insects, however, in shedding their skin followed by the growth of new skin, which is called molting. It is quite fascinating that biological control of pests that affect the primary production of crops is carried out by spiders, as spiders can devour extraordinary numbers of agricultural pests (insects that affect crop plants in cultivation and plantation) but do not affect the crops themselves.

According to recent studies on the species diversity of spiders, 40,000–45,000 species of spiders are dispersed throughout the world (Coddington and Levi, 2003; and Spagna and Gillespie, 2008; Wessel, 2018). The species diversity of spiders falls under criteria for shape, size, and habitat that have caused zoologists and naturalists to recognize spiders as the seventh-ranked species globally for species diversity. Spiders are present everywhere and in nearly all habitats on six continents (excluding Antarctica). Earlier it was observed by arachnologists that polar regions and oceans are not suitable habitats, but mountaineers on Everest expeditions have observed the existence of jumping spiders, and contemporary oceanographic expeditions have affirmed that species can be found within coral reef ecosystems and rock crevices at the intertidal zone of the ocean's edge. These recent field observations on spider biodiversity do not support the erstwhile observations. From a geological and paleogeographical perspective, along with Wegner's tectonic plate movement hypothesis and continental drift theory, there was only one supercontinent, "Pangea," that existed in the middle of the superocean "Panthalassa" during the prehistoric age, which broke into several continental plates that floated in the ancient oceans, drifting apart for millions of years. Those continental plates drifted with animals and plants on their backs and migrated to different climatic regimes after being dispersed throughout the ocean bodies. The intermittent drifting of the continental shield, until it stabilized, forced the existing animals and plants inhabiting it to diversify further. As a tenable instance, geologists have revealed that contemporary Australia was part of the southern supercontinent Gondwana around 150 million years ago. The breakaway part of Gondwana created an isolated environment for the spiders that had been part of it. Over time, the origin and evolution of new spider species and infraspecies happened independently. Comparative phylogenetic studies of ancestral species of plants and animals in South America and Africa, recently evolved species in Australia, and the observation of distant phylogenetic relations among them, indicate prehistoric migrations of spiders in parallel with the continental drift hypothesis. The wind-assisted dispersal of spiders from one ecological region to a distant one is called ballooning. Arachnologists have observed that some small-sized and juvenile spiders are blown away with a piece of silk thread under strong winds and travel long distances to reach distant lands across the seas until they are caught up by obstacles or the blowing wind or storm ceases. This ballooning technique physically resembles the aerial plankton hypothesis promulgated by Majerus (2009) to define the long-range dispersal of the peppered moth in Britain during the Industrial Revolution. Arachnologists observed that long-range dispersal of orb weavers and wolf spiders successfully resulted from the dispersal of spiders in the new habitat driven by "ballooning."

Scientists have observed that most of the time, spiders thrive in dry, terrestrial habitats mainly on land, often on trees, and in human settlements, and the diversity of spiders can normally be found in tropical and subtropical ecoclimatic zones rather than in high altitudinal temperate and alpine ecoclimatic regions. Although a few arachnid species are found in the ecotone of land and water bodies, the diversity of spiders in wetlands, meadows, and bodies of water such as lakes and ponds is considerably low. Arachnologists have noticed that the same spider species can live longer in captive environments, mainly human environments or abandoned human settlements, as long as the concerned spiders can weave silk webs to trap enough food, but they have shorter life spans in the wild due to a high level of interspecific competition that often marginalizes the availability of food.

No matter whether housekeeping is performed regularly, spiders tend to make their own space in human settlements—for example, in closets, storage, barns, or ceilings. During the regime of the Fourth Industrial Revolution, as anthropogenic actions are directly responsible for wild habitat destruction, particular species of spiders would have been forced to migrate to some uncharted locations where they had never lived, and they occupied those habitats if selection pressures favored them to thrive under the microclimatic conditions of occupying such habitats. Often, spider species have migrated in the shipping containers of food crops, plant specimens, and forest products as a result of flawed enforcement of animal quarantine procedures across countries. Several indirect anthropogenic activities that cause climate change, prolonged drought, unexpected floods, global warming, sea-level changes, frequent occurrences of earthquakes, and redundant natural catastrophes such as El Niño often drive spiders to take refuge in isolated refugia in caves, caverns, isolated mountaintops (which also act as islands), or islands. According to the island, or insular, biogeography theory of MacArthur and Wilson (2001 repr.ed.), the isolated population of any biological species has the opportunity to become an evolved endemic species, recognized as "island endemics" and in isolation. The evolution of new endemic spiders if the particular species in isolation has a fairly substantial population size. Small spider populations in isolation often suffer from genetic erosion, recognized as the "bottleneck effect," which may lead to the evolution of new variations that adapt to such a microhabitat or specialized habitat, or the species might be forced to the verge of extinction. But there is another reality, that the evolution of new species does not necessarily ensure a new lease on life to an organism's group, as a species that evolved and survived in a fragile habitat or in isolated refugia could not disperse beyond the specialized habitat and unique microclimatic environment. But the evolutionary journey of life is not a journey on the highway, as evolutionary success always goes through the labyrinth of space and time to ensure the shift in the paradigm of life from the ancestral to derived form.

The structural pattern of the body of the members belonging to Arachnida is simple and helps spiders move with hydraulic pressure-driven movement that saves energy and is required for movement, capturing prey, and the reproductive cycle. The main difference between the members of insects and arachnids is clearly in the body structure. Whereas the insect body has three distinct parts—the head, thorax, and abdomen—arachnids have two main body parts, the cephalothorax and abdomen. According to Jo-Anne Nina Sewlal, the renowned arachnologist at the University of the West Indies and Trinidad, two main body parts of the spider are the prosoma or cephalothorax (a fused structure of head and thorax) and the opisthosoma or abdomen (Szalay, 2014). The cephalothorax or prosoma contains the important organs and limbs of the spiderlike eyes, mouth, and legs. The spider's jaw is an important part of the mouth called "chelicerae," a fang-like structure that helps it catch prey by injecting venom to liquefy the solid food. Just behind the jaw, labium, and labarum, two important mouthparts help the spider take in food. Between the chelicerae and the first pair of legs are small, insipient leglike structures called "pedipalps" that work like sensory organs to help catch prey. According to scientists, these pedipalps play an important role in the mating process; the tips of the pedipalps of male spiders are enlarged relative to their female counterparts and act to direct the sperm discharged by the male to go through the genitals of the female spider (Szalay, 2014). The abdomen or opisthosoma is one of the important body parts of the spider, as it contains lungs, digestive tracts, reproductive organs, and "spinnerets," and spinnerets help spiders produce

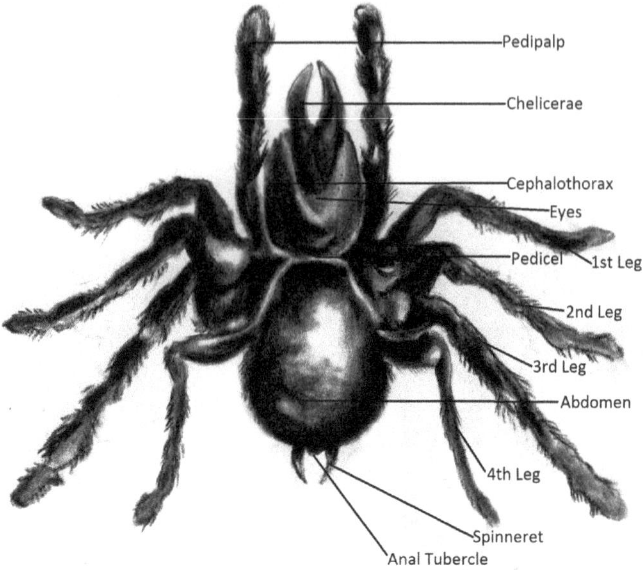

Figure 3.1 Body parts of the spider.

silk to weave webs with silk threads. The shedding of the exoskeleton of its outer body happens regularly in an annual cycle. The diagrammatic sketch of the body parts of a spider is presented in Fig. 3.1.

Preliminary studies on spiders have revealed some 45,000 to 50,000 spider species in the world (some arachnologists even consider the number to be around 180,000 species). Some unique and worrisome-looking species of spiders or their poisonous members are briefly discussed herein (Tavolacci, 2003; Platnick, 2011; World Spider Catalog, 2020):

1. Banana spider: The Banana spider is also known as the "Golden silk orb-weaver" and is scientifically known as *Nephila clavipes* of the Tetragnathidae family. The size of the female Banana spider ranges from 4.9 to 5.2 cm, whereas its male counterpart ranges from 2.5 to 2.6 cm. This bright-looking (with a golden orange bicolor combination) spider is present in the tropical region of the southern United States, and this apparently scary species does not have poisonous venom (World Spider Catalog, 2020). It weaves a web the color of silk spun with a golden tinge, which visually stands out from other spider webs.
2. Black widow spider: The "Black widow spider" generally comprises 31 species of the genus *Latrodectus*, of which the three most venomous

species are scientifically known as western black widow or *Latrodectus hesperus*; northern black widow or *L. variolus*; and southern black widow or *L. mactans*, belonging to the Theridiidae family. These species are considered some of the most venomous spiders in North America, the sting of which is sometimes fatal to human beings (Szalay, 2014). The Redback spider or *Latrodectus hasseltii*, which is also recognized as the Australian black widow spider, is one of the most venomous arachnids originating in the desert regions of Southern Australia (Whyte and Anderson, 2017). The size of the Black widow spider ranges from 3 to 13 mm. The Black widow spider is well distributed over temperate regions of Africa, Australia, Asia, North America, and Southern Europe.

3. Brazilian wandering spider: Though a group of arachnologists recognized two species of the genus *Phoneutria*, *P. fera* and *P. nigriventer*, as the Brazilian wandering spider, another group of arachnid specialists considered the entire genus—i.e., eight species of *Phoneutria* are considered to be Brazilian wandering spiders (Clarke, 2014; World Spider Catalog, 2020). The Brazilian wandering spider's size varies from 18 to 49 mm. Its legs, stretched from the distinguishing long stripe in its dorsal palp, and the thin line of the stripe in the anterior-posterior part of the dorsal carapaces, make it visually recognizable. Most species of *Phoneutria* can be found within the tropical rainforest ecosystems of South America (Martins and Bertani, 2007). It is one of the deadliest venomous spiders in the world and kills a wide range of prey: crickets, millipedes, lizards, mice, snakes, frogs, etc. The sting of this deadly spider venom causes serious ailments, such as loss of muscular function, paralysis, and asphyxiation, in human beings. Several biochemical/clinical experiments and a biological assay on the spider's venom, specifically known as Tx2-6, are ongoing to cure erectile dysfunction (National Geographic Society Newsroom, 2012).
4. Brown recluse spider: The Brown recluse spider is popularly recognized as *Loxosceles laeta*, but it comprises around 139 species belonging to the Sicariidae family. Its size varies from 6 to 20 mm. The most venomous six-eyed species of *Loxosleles* have been found distributed within warmer locations of North America, South America, the Caribbean, Central America, Asia, Africa, Finland, and Australia (World Spider Catalog, 2020). It is dark brown, with no distinguishing color pattern, and is characterized by its structural similarity to the shape of a violin, which is why it is sometimes called the Violin spider. It can survive in a dormant state for a long time without consuming food and water. Its

bite causes severe physical and physiological ailments in other animals, such as rupturing red blood cells, damaging tissues, and damaging skin tissues on the body surface. Thus far, scientists have been unable to find an antivenom to the venom injected by this spider into its prey (Magalhaes et al., 2017).

5. Camel spider: The "Camel spider" is scientifically known as *Galeodes arabs* and belongs to the Galeodidae family under the order Solifugae, or the Sun spider, underclass Arachnida (Levin, 2013). The size of the Camel spider ranges from 14 to 15 cm. Solifugae has around 150 genera and 1000 species (Barnes, 1982). Though it is popularly known as the Camel spider, it doesn't belong to the order Araneae (spiders); rather, it is a transitional form between the order Scorpiones (scorpions) and spider, which has taxonomically been treated under the common name Sun spider or Solifugae (Cloudsley-Thompson, 1961). Camel spiders are similar to the spider except for the presence of a pair of chelicerae, commonly known as a pair of jaws, and the Camel spider has 10 legs instead of the 8 found in most members belonging to the order Araneae (Szalay, 2014). Though most Camel spiders are found in the desert regions of Middle West Asia and West Africa, these are found in the southwestern United States and Mexico (Punzo, 1998). During the Gulf War in 1990 and the US attack on Iraq in 2003, the Camel spider drew media attention for its nocturnal movement (such as running at 30 mph and jumping 3 feet during night time or making hissing noises and chasing soldiers in the desert) and such observation has been made by the US soldiers and media observers, rather than studied by field ecologists or arachnologists. It has been reported to kill and eat insects, birds, lizards, and rodents using its jaws. So far, there are no reports of the nonvenomous camel spiders attacking human beings and creating human ailments, but its bite on some occasions has been reported as painful (Szalay, 2014c).
6. Diving bell spider: The "Diving bell spider," or Water spider, scientifically known as *Argyroneta aquatica*, belongs to the family of Dictynidae and spends almost its entire life underwater. It is dark brown in the cephalothorax and dark gray in the abdomen (Schutz et al., 2007). An aquatic, it performs its major physical and physiological functions underwater to complete its life cycle except on a few occasions when it comes to the water's surface for extra oxygen and predation (Yong, 2011). Though some species of the genus Argyroneta are semiaquatic, this Water spider is found in the freshwater ecosystems of Europe,

Central Asia, and East Asia. Most male terrestrial spiders are smaller than their female counterparts, but in the case of the Diving bell spider, males are usually larger and more proactive than females (Semour and Hetz, 2011). The size of the male Water spider varies from 10 to 15 mm, whereas the size of the female varies from 8 to 12 mm. So the hypothetical advantage of sexual dimorphism, that females can lay an enormous number of eggs, is not applicable to these particular situations. Arachnologists have construed that the ecological environment of an aquatic ecosystem led these spiders to overcome water resistance to maintain buoyancy, better diving potential, fast movement to catch prey, etc. These factors rendered the males more energetic, more proactive, and larger than females (Schutz and Taborsky, 2003). Earlier, arachnologists presumed the Diving bell spider could consume enough oxygen by making a few trips to the surface, but recent experimental findings revealed that this aquatic spider weaves webs in submerged conditions by tying the web to submerged aquatic plants and utilized the underside of the submerged web as an oxygen storage location in the form of tiny bubbles, which they have further used in long-term breathing (Kaplan, 2011). An instance of sexual dimorphism witnessed in the Common mouse spider (*Scotophaeus blackwalli*) is presented in the diagrammatic sketch of Fig. 3.2.

7. Hobo spider: The hobo spider, scientifically known as *Eratigena agrestis*, is light brown and belongs to the family Agelenidae. It is normally found in the woods and forest ecosystems of Europe, Central Asia, and the Pacific coast of North America (Hadley, 2019). The size of the female Hobo spider ranges from 11 to 15 mm, whereas the male ranges from 8 to 11 mm; it is also called the Funnel-web spider, as it builds a funnel-shaped silk web that it uses to trap insect prey (Bolzern et al., 2013). The Hobo spider is distinct from other members of Agelenidae in its absence of bright color bands at the leg joints, and its abdomen has a V-shaped pattern (Vetter and Visscher, 2001). Whether the Hobo spider is venomous is a matter of debate to arachnologists, as the Centers for Disease Control in Atlanta, USA, reported cases of necrosis in some people in 1990, though they did not place the Hobo spider under the category of venomous spider in 2017 (Crawford, 2015). In Canada, there have been no reports of necrosis or skin sensitization issues due to Hobo spider bites (Blaszczak-Boxe, 2014).
8. Jumping spider: "Jumping spiders" are a group of spiders that belong to 600 genera and 6000 species under the family Salticidae, which is the

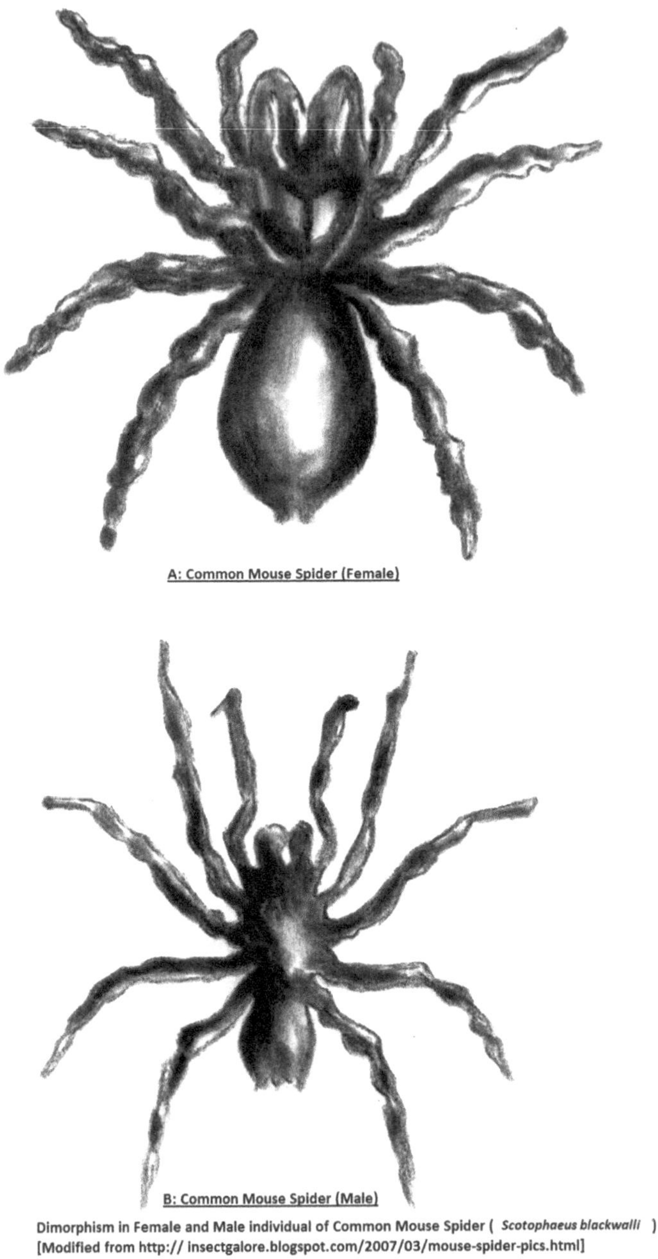

Figure 3.2 Sexual dimorphism in female and male individuals of Common mouse spider (*Scotophaeus blackwallii*).

largest spider family and encompasses 13% of all spider species. It is characterized by its rectangular flat face with four pairs of eyes—a pair of large-sized anterior–median eyes (functional in regular light), and smaller anterior lateral, posterior lateral, and posterior median eyes (functional under UV light) (Harland and Jackson, 2000). It uses two different respiratory organs book lungs and tracheal systems to support bimodal respiration (Jocque and Dippenaar-Schoeman, 2007). The size of the Jumping spider varies from 1 to 25 mm, and they are found in almost all terrestrial ecosystems, ranging from the Gobi Desert to the high-altitude rocky clefts of the Everest mountain ranges (Wanless, 1975). It uses four pairs of legs to catch prey—the hydraulic pressure driven by alteration of the pressure in hemolymph triggers the posterior legs used for jumping in food intake or movement. *Hyllus giganteus*, the largest Jumping spider (around 25 mm), can make a giant leap and utilize the silk thread as a safety line that helps it stabilize its landing (BBC Earth Unplugged, 2016). Almost all spider species are carnivores, except one species, *Bagheera kiplingi*, that survives on plant matter and nectaries (Milius, 2008). Male jumping spiders attract female spiders for courtship by exposing their iridescent colorful body parts under UV light reflection and zigzag movement, called the "courtship dance." Some males make drumbeat-like auditory signals to attract females (Elias et al., 2003; Bulbert et al., 2015).

9. Tarantula: Tarantulas are large hairy spiders belonging to 116–123 genera of 850–1000 species under the family Theraphosidae (World Spider Catalog, 2020). The name tarantula was picked from a Mediterranean species, *Lycosa tarantula.* Though most tarantulas in North America are dark brown, in other parts of the world, their color varies from dark blue-black to black-and-white striped (Szalay, 2014). The size of the body of the Tarantula spider ranges from 2.5 to 10 cm in length, and the size of the leg varies from 8 to 30 cm. Tarantulas are well dispersed in trees (arboreal) and on the ground (terrestrial) of forest ecosystems throughout Asia, Africa, North America, South America, Central America, and Mediterranean Europe. Pink toe or *Avicularia avicularia* is one of the most fascinating tarantulas and can be found in the Caribbean region. It has a strong physiological resemblance to the arthropods in molting the exoskeleton (McLean et al., 2018). Sexual dimorphism has been found in these members. It is one of the venomous spiders that kills large numbers of arthropods (insects) and tarantulas as prey; moreover, it kills and eats a number of small vertebrates, such as snakes, lizards, mice, and bats. The "Goliath birdeater spider" belongs to the tarantula group, and arachnologists consider it one of the largest spiders in the world.

10. Wolf spider: The Wolf spider is one of the top three categories of spiders with perfect vision and the ability to move fast and catch prey. It seems they have muscular locomotory organs (wolfspiders.org, 2018). The size of the Wolf spider ranges from 10 to 35 mm. There are around 210 species of two predominant genera: genus *Hogna*, with 200 species, and genus *Hoggicosa*, with 10 species, distributed throughout Asia, Africa, North America, the Caribbean, South America, Australia, and Oceania. *Hogna carolinensis* is one of the renowned wolf spiders in North America (Piacentini and Ramirez, 2019). The Wolf spider has mild venom, though the sting of Australian and Oceanian wolf spiders causes necrotic wounds (World Spider Catalog, 2020).

The "Giant bird eating spider" or "Goliath birdeater spider" is considered the largest spider in the world, though some arachnid specialists consider the "huntsman spider" the largest if leg span is considered. Conversely, *Patu digua*, which is endemic in Colombia, is considered the smallest spider in the world. Though arachnologists have meticulously studied a diverse group of arachnids regarding their habits, habitat life cycle, mode of reproduction, etc. with the kind of intensive investigations to help scientists provide several previously unknown facts and figures, some questions have yet to be answered and remain unsolved, such as the reproductive biology of the spiders. Arachnologists have observed carefully that most of the time, female spiders are structurally larger than their male counterparts. On some occasions, mature male spiders eject sperm in locations that attract their female counterparts to fertilize their eggs with that sperm, but on most occasions, the female spider physically mates with their male counterpart and becomes inseminated, and at the end of the courtship, the larger female devours her male partner. This sexual cannibalism in arachnids remains a mystery for scientists. Successful mating of the female spider results in them laying a tremendous number of fertilized eggs, sometimes around 3000 (Aisenberg et al., 2011). Many arachnologists have tried to hypothesize reasons for the mysterious cannibalistic behavioral trait of female spiders by putting forth an opinion that most likely female spiders fulfill their demand for protein by eating the males, which might help them maximize production by being able to lay an enormous number of fertilized eggs, and that kind of reproductive success made the spider one of the diverse groups of invertebrates in the world in terms of species diversity that strived for the successful occupation of the diverse range of ecological niches, across the biosphere (Lomborg and Toft, 2009). By studying a comparative account of the ratio between male and female members of a particular species, arachnologists noticed that in most of the spider community, there was a domination of female members in terms of numbers in comparison with the lower number of male members, and scientists contemplated further that mysterious postmating devouring of male members by their female counterparts is responsible for such unequal ratio of male and female individuals in a spider community.

Though most spiders survive in the wild habitat for a short period varying from 2 to 3 years, scientists have observed that in captivity, mainly in close and interior microenvironments, they can survive for more than 10 years. Under physical and physiological adaptations, these small-sized arthropods spread throughout the world, occupy diverse types of habitats, survive in a wider range of environmental conditions (from the Amazon rainforest to Saharan deserts to Arctic tundra in circumpolar regions). Though the size of spiders is diverse, the content of venom or their poison-induced physical immobilization or paralyzing potential has nothing to do with size, which indicates further that big-sized spiders are not necessarily poisonous (the stereotypic perception being that they contain venom); rather, human beings stung by small-sized spiders became paralyzed or die because of the extent of toxicity of the venom infused into their bloodstream. In this aspect, it should be mentioned that the "Banana spider" from Brazil is the most poisonous spider in the world (Tavolacci, 2003). Most spiders indeed have venom, but qualitative research has determined that incidents of spider venom—induced sickness or death among human communities are not so vivid, as stereotypic perceptions against spiders have prevailed in our society. "Arachnophobia" is the most common psychological fear perceived by society and has normally registered 8%—12% of men and 48%—52% of women (Vetter and Visscher, 2001).

The members of Arachnida are well adapted to changes in space or environmental conditions that have helped this creature maintain a high rate of species diversity and occupy a vast range of ecological niches in a wide range of ecoclimatic regions throughout the world. They have biological safeguards (being venomous) against potential predators, so it seems the high extent of species diversity would supposedly remain unchanged. But the increasing pace of anthropogenic interference in wild habitats leads to habitat fragmentation that is indirectly responsible for the shrinkage of population size and the alleviated level of genetic erosion that tends to stall the diversification of species diversity; on the other hand, depletion of wild habitats to human settlement or cultivated land, and the increased level of insecticide use, directly exterminates the normal population of spiders and result in downsizing the species diversity of arachnids (Tavolacci, 2003). To ensure their survival, spiders often struggle to survive with minimal food intake. Their acts of predation are precisely witnessed when they contact the tiny insects that can be found entangled in their webs. Eventually, the spider silks they make come out from their secreting glands. After an intensive investigation on the biochemical properties of the webs weaved by spiders, biochemists have concluded that the silk weaved by the spider is one of the most sturdy materials in the world, which has intrigued chemical engineers and fiber technologists who have tried to recreate it using all available technologies but have failed (Spagna and Gillespie, 2008). Scientists have observed that the sturdiness of the spider web helps keep dust particles away from the web. Web-spinning by a spider is a continuous process that involves converting the old web into a ball before weaving a new one, and many species eat these balls. Spiders are small-sized predators and literally carnivores, as they collect their nourishment from other insects by sucking out all liquid nutrients after injecting venom into prey to convert their bodies from semisolid into liquid form. As spiders have neither evolved calcified teeth nor a well-developed digestive system like that of vertebrates, none of them can chew and digest solid food, and thus, they depend on their venom to disintegrate solid food into liquid form, and the spiders suck that liquid form of food for nourishment. Though almost 99.9% of spiders are carnivorous or insectivorous, one species is herbivorous (i.e., *Bagheera kiplingi*) (Jocque and Dippenaar-Schoeman, 2007). It's quite interesting that most spiders have poor vision and reach their food source using touch sensors in their limb tips. As soon as the prey becomes entangled in the web, the spider's touch sensors can sense the location of the food source by sensing the quake drifting through the web; it can then inject venom into the prey to immobilize/kill it to eat it further.

Though these creatures are small and slow-moving (mainly due to four pairs of legs that are not well adapted to maintaining body balance and running fast), they are readily protected from the predation pressures of the other big-sized carnivorous and omnivorous animals, as they were evolved with special features such as potential production of venom, a biological safeguard that has been injected into predators; if those predators get close to the spider for a predatory attack, they become paralyzed or are killed by the poisonous effects of the venom (Jocque and Dippenaar-Schoeman, 2007). Still, spiders are devoured by many predators, including birds, frogs, wasps, lizards, and monkeys. Spiders are devoid of any muscles attached to their locomotory organs, such as jointed legs. Hence, the combination of short-sightedness and hydraulic power-driven movement, steered by jointed legs without any muscular tissues, results in its slow movement. This survival technique of spiders has made most human beings frightful of it. The jointed-leg-driven movement of the spider is quite interesting at the time of movement—four legs of any spider will be touching the ground, while the remaining four legs will be above the ground. Most spiders are slow-moving, but there are exceptions. Autecologists have noticed that the "Jumping spider" has the jumping potential to traverse more than 50 times its body length, which happens mainly because of different blood pressure in the hind limbs of its body's posterior that helps exert pressure to jump forward such a distance at a stretch (Platnick, 2011). As far as distinct identification of the diverse varieties of the spider species is concerned, the contrasting colorful existence of many spider species has utilized the colorful surfaces of their bodies to act as a safeguard that protects them from predatory attacks by camouflaging them with adjacent living creatures to create a scary look that makes predators stay away. According to most zoologists and arachnid experts, there is a color-coding system that further helps new researchers and scientific workers become familiar with the identities of spiders and determine which ones are truly poisonous and which ones seem poisonous but are not.

As long as we are concerned about the stability of our biosphere and maintaining the equilibrium of biodiversity on Earth, we need to agree that the spider is a part of biological diversity as long as its population dynamics remain uninterrupted. Staying away from the biological importance of this creature, even as we consider its ecological importance to the biological control of agricultural pests, the commercial craving of spider venom in the pharmaceutical industry (for production of cardioprotective and reproductive health medicines); agricultural industries (as biochemical pest control products); and in fiber–textile industries (as transgenic silk fibers) should not be overlooked. To exploit the untapped biological treasure trove of arachnid sources, we must ensure the sustainability of the population of spiders along with their species diversity. Vast wild forest ecosystems have been destroyed worldwide since the beginning of the Industrial Revolution. Such destruction has been observed in various parts of the world, including clearing bushlands to transform them into human settlements in Australia and clear-felling of lumber-yielding trees in Southeast Asia and South America; both events pushed spider habitats to the verge of drastic depletion in population size. Habitat conservation and restoration help restore the ecosystem to make it pristine and rehabilitate the shrunken and depleted population of spiders in their natural habitats. Spider conservation should not be considered a utopian idea of radical conservationists but rather part of the broader conservation mission of reforming the tilted equilibrium of biodiversity on Earth and an ultimate remedial measure to correct massive anthropogenic misdeeds such as ever-increasing carbon emissions-induced global warming, ozone layer depletion in the stratosphere, the meltdown of glacier ice, and the gradual and intermittent rise of sea levels.

References

Aisenberg, A., Costa, F.G., Gonzalez, M., 2011. Male sexual cannibalism in a sand-dwelling wolf spider with sex role reversal. Biol. J. Linn. Soc. 103 (1). https://doi.org/10.1111/j.1095-8312.2011.01631.x.

Barnes, R.D., 1982. Invertebrate Zoology. Holt-Saunders International, Philadelphia, PA, p. 613, 614.

BBC Earth Unplugged, 2016. Watch the World's Biggest Jumping Spider Make a Leap. http://www.bbc.com/earth/story/20160129-watch-the- worlds-biggest-jumping-spider-make-a-dramatic-leap.

Bulbert, M.W., O'Hanlon, J.C., Zappettini, S., Zhang, S., Daiquin, L., 2015. Sexually selected UV signals in the tropical ornate jumping spider, Cosmophasis umbratica may incur costs from predation. Ecol. Evol. 5 (4), 914–920. https://doi.org/10.1002/ece3.1419.

Blaszczak-Boxe, A., 2014. Venomous Hobo Spider Bites May Be Not So Toxic after All. https://www.livescience.com/44907-hobo-spider-bites.html.

Bolzern, A., Burckhardt, D., Hänggi, A., 2013. Phylogeny and taxonomy of European funnel-web spiders of the Tegenaria-Malthonica complex (araneae: agelenidae) based upon morphological and molecular data. Zool. J. Linn. Soc. 168, 723–848. https://doi.org/10.1111/zoj.12040.

Campbell, D., 2016. Aristotle's on the Heavens. https://www.ancient.eu/article/959/aristotles-on-the-heavens/.

Clarke, D., 2014. Venomous Spider Found in Waitrose Shopping 'Beautiful but Aggressive'. The Guardian. Retrieved June 22, 2015.

Cloudsley-Thompson, J.L., 1961. Some aspects of the physiology and behaviour of Galeodes arabs. Entomol. Exp. Appl. 4 (4). https://doi.org/10.1111/j.1570-7458.1961.tb02140.x.

Coddington, J., Levi, H., 2003. Systematics and evolution of spiders (araneae). Annu. Rev. Ecol. Systemat. 22, 565–592. https://doi.org/10.1146/annurev.es.22.110191.003025.

Corrigan, J., Denny, F., Jaffee, M.S., Eire, C., 2011. Jews, Christians, Muslims: A Comparative Introduction to Monotheistic Religions. Routledge, p. 408.

Crawford, R.L., 2015. Myths About "Dangerous" Spiders. http://www.washington. edu/burkemuseum/spidermyth/myths/hobo.html.

Dalal, R., 2010. Hinduism: An Alphabetical Guide. Penguin Books India, p. 483.

Elias, D., Mason, A., Maddison, W., Hoy, R., 2003. Seismic signals in a courting male jumping spider (araneae: salticidae). J. Exp. Biol. 206, 4029–4039.

Hadley, D., 2019. Hobo Spider (*Tegenaria agrestis*): Habits and Traits of Hobo Spiders. https://www.thoughtco.com/hobo-spider-tegenaria- agrestis-1968553.

Harland, D., Jackson, R., 2000. Eight-legged cats' and how they see - a review of recent research on jumping spiders (araneae: salticidae). CIMBEBASIA 16, 231–240.

Horowitz, W., 1998. Mesopotamian Cosmic Geography. The Pennsylvania State University, p. 418.

Ibn 'Arabi, M., Al'Arabi, I., 2008. Divine Sayings: 101 Hadith Qudsi: The Mishkat Al-Anwar of Ibn 'Arabi. (Translated by Stephen Hirtenstein and Martin Notcutt). Anqa Publishing, p. 120.

Jansma, R., Jain, S.R., 2006. Introduction to Jainism. Prakrit Bharti Academy, p. 236.

Jocque, R., Dippenaar-Schoeman, A.S., 2007. Spider Families of the World. Royal Museum for Central Africa. Tervuren, p. 338.

Kaplan, R.L., 2011. Spider-man in love: a psychoanalytic interpretation. J. Popular Cult. 44 (2). https://doi.org/10.1111/j.1540-5931.2011.00833.x.

Levin, S. (Ed.), 2013. Encyclopedia of Biodiversity, second ed. Academic Press, p. 5504.

Lomborg, J.P., Toft, S., 2009. Nutritional enrichment increases courtship intensity and improves mating success in male spiders. Behav. Ecol. 20 (4), 700–708. https://doi.org/10.1093/beheco/arp044.

MacArthur, R.H., Wilson, E.O., 2001. The Theory of Island Biogeography. Princeton University Press, p. 224.

Magalhaes, I., Brescovit, A., Santos, A., 2017. Phylogeny of sicariidae spiders (araneae: haplogynae), with a monograph on neotropical sicarius. Zool. J. Linn. Soc. 179, 767–864. https://doi.org/10.1111/zoj.12442.

Majerus, M.E.N., 2009. Industrial melanism in the peppered moth, *Biston betularia*: an excellent teaching example of Darwinian evolution in action. Evol. Ed. Outreach 2, 63–74. https://doi.org/10.1007/s12052-008- 0107-y.

Martins, R., Bertani, R., 2007. The non-Amazonian species of the Brazilian wandering spiders of the genus *Phoneutria perty*, 1833 (araneae: ctenidae), with the description of A new species. Zootaxa 1526, 1–36. https://doi.org/10.5281/zenodo.177546.

McLean, C.J., Garwood, R.J., Brassey, C.A., 2018. Sexual dimorphism in the arachnid orders. PeerJ 6, e5751. https://doi.org/10.7717/peerj.5751.

Milius, S., 2008. Life: spider eats trees, not bugs: small jumping species steals lunch from ants. Sci. News 174 (5). https://doi.org/10.1002/scin.2008.5591740514.

National Geographic Society Newsroom., 2012. Venom from the Banana Spider Could Be the New Viagra. https://blog.nationalgeographic.org/2012/09/10/venom-from-the-banana-spider-could-be-the-new-viagra/.

Pennisi, E., 2018. Part Spider, Part Scorpion Creature Captured in Amber. https://www.sciencemag.org/news/2018/02/part-spider-part-scorpion-creature-captured-amber.

Piacentini, L.N., Ramírez, M.J., 2019. Hunting the wolf: a molecular phylogeny of the wolf spiders (araneae, lycosidae). Mol. Phylogenet. Evol. 136, 227–240. https://doi.org/10.1016/j.ympev.2019.04.004.

Platnick, N.I., 2011. The World Spider Catalogue, Version 12.0. American Museum of Natural History. Available from: http://research.amnh.org/iz/spiders/catalog/INTRO3.html. (Accessed 6 September 2011).

Punzo, F., 1998. The Biology of Camel-Spiders. Springer Science, Business Media, New York, p. 301.

Schuetz, D., Taborsky, M., 2003. Adaptations to an aquatic life may be responsible for the reversed sexual size dimorphism in the water spider *Argyroneta aquatica*. Evol. Ecol. Res. 5, 105–117.

Schütz, D., Taborsky, M., Drapela, T., 2007. Air bells of water spiders are an extended phenotype modified in response to gas composition. J. Exp. Zool. 307A, 549–555.
Semour, S.R., Hetz, S.K., 2011. The diving bell and the spider: the physical gill of *Argyroneta aquatica*. J. Exp. Biol. 214, 2175–2181. https://doi.org/10.1242/jeb.056093.
Spagna, J., Gillespie, R., 2008. More data, fewer shifts: molecular insights into the evolution of the spinning apparatus in non-orb-weaving spiders. Mol. Phylogenet. Evol. 46, 347–368.
Szalay, J., 2014a. Types of Spiders & Spider Facts. https://www.livescience.com/22122-types-of-spiders.html.
Szalay, J., 2014b. Black Widow Spider Facts. https://www.livescience.com/39919-black-widow-spiders.html.
Szalay, J., 2014c. Camel Spiders: Facts and Myths. https://www.livescience.com/40025-camel-spiders-facts.html.
Szalay, J., 2014d. Tarantula Facts. https://www.livescience.Com/39963-tarantula.html.
Tavolacci, J., 2003. Insects and Spiders of the World, vol. 11. Cavendish Square, p. 704.
Vetter, R.S., Visscher, P.K., 2001. Bites and Stings of Medically Important Venomous Arthropods. Department of Entomology, University of California-Riverside.
Wanless, F.R., 1975. Spiders of the family salticidae from the upper slopes of Everest and Makalu. Bull. Brit. Arach. Soc. 3 (5), 132–136.
Wessel, L., 2018. Sticky Science: The Evolution of Spider Webs. https://www.scientificamerican.com/article/sticky-science-the-evolution-of-spider-webs/.
Whyte, R., Anderson, G., 2017. A Field Guide to Spiders of Australia. https://doi.org/10.1071/9780643107083.
Wolf S. Arachnidism & Pain, Habitat, Diet and Cannibalism. wolfspiders.org.
World Spider Catalog, 2020. World Spider Catalog. Version 21.5. Natural History Museum Bern. Online at: http://wsc.nmbe.ch.
Yong, E., 2011. The Diving Bell and the Spider. https://www.nationalgeographic.com/science/phenomena/2011/06/09/the-diving-bell-and-the-spider/.

CHAPTER FOUR

Spider life span: Web-weaving with a wispy thread of silk

Abstract

Conspicuous morphoanatomical and biobehavioral studies of arachnids have helped scientists, particularly evolutionary biologists, recognize that their dimorphism is integrated into their gender orientation. Female spiders are larger, physically stronger, more aggressive, and more venomous than their male counterparts. Scientists have also noticed that the unique completion of the spider's sexual cycle ends with the devouring of the male spider by its female counterpart, which is an act of cannibalism. From the meticulous way that spiders weave webs, to its monochromatic vision-assisted hydraulic movements, to its means of catching prey by immobilizing it with venom injection, the entire physical and physiological maneuvering of arachnids has enabled these unique, evolutionarily primitive creatures to survive in this regime of climate change since their origin some 350 years ago.

Keywords: Cannibalism; Reproductive cycle; Sexual dimorphisms; Spiders.

For me, each day begins and ends with wanting to learn a little more about the secrets of spider silk.

Cheryl Hayashi

Although the morphology and anatomy of the spider might not appear unique at first glance, the morphological variation in size and color presents a perplexing identity and functions. The diverse colors and body shapes, hair-like structures, etc., are used by spiders as natural camouflage in wild habitats to save themselves and averting predatory attacks (Foelix, 2011). But a closer look helps one notice the intricate details and determine the identity of the species and the infraspecific level and provide ideas about its general physical, physiological, and biobehavioral features, which are fascinating indeed (Walcott, 1969; Barth, 2002). Structural dimorphism in spiders generally prevails across a diverse group of species in spiders, and females are usually much larger than their male counterparts. The larger-sized females are generally more aggressive, stronger, and more venomous than males. It is quite something to witness the cannibalism of arachnids, which is integral

Spider Evolution
ISBN: 978-0-323-90041-6
https://doi.org/10.1016/B978-0-323-90041-6.00012-7

to successfully completing their sexual cycle, as most females devour their male counterparts after sexual mating is completed.

All arachnids have four pairs of joint legs, laterally the side of the body, and they are devoid of any muscular attachments. Naturally, those legs are not biomechanically convenient for maintaining body balance and are thus driven by hydraulic pressure that is triggered by the differential pressure of body fluid as it flows from the front to the back of the spider's body (Altherr, 2019). Most spiders have four pairs of eyes (some have three pairs), but even so, they are typically short-sighted with a monochromatic vision not suited to see predators or prey from long distances. However, it helps them steer a particular direction to offset the lack of balance in its hydraulic power-driven locomotion (Pomeroy, 2013). Overall, the combination of legs and eyes allows ease of movement and energy conservation at the same time. The feeding mechanism and procurement of energy of this unique group of invertebrates are quite interesting. Arachnologists observe that the central nervous system of spiders very strong to sense the appearance of predators or prey to hide or becoming aggressive (Walcott, 1969; Fabian-Fine et al., 1999) using the touch and biochemical sensors and the food, sending signals to mating partners, etc. (Barth, 2002).

However, spiders are not capable of eating solid food (due to lack of jaws and teeth), so they use their fangs to inject venoms into the bodies of predators, which is strong to kill or immobilize the prey and the solid and semisolid into the liquid fluidlike structure that has been sucked to meet the demand of the metabolic energy to survive (Walter et al., 2017). Meticulous investigations done by arachnologists helped elucidate the overall impression dispersal of spiders all over the world in terms of species and ecodiversity almost everywhere (except Antarctica) in the world enabled prehistoric invertebrates to survive for a long time with its limitations (as they are devoid of advanced features like invertebrates) and their successful thriving happens due to its intermittent changing spree in terms of morphological and anatomical characteristics in the backdrop of the constant changes of dynamic environmental conditions in its ambiance (Walcott, 1969; Foelix, 2011).

However, the morphological and anatomical characteristic features of spiders its skin tissues, respiratory systems (having tracheal tubes and book lungs), digestive systems (having the muscular, pumping stomach to pump liquid food into the food pipe or esophagus and further move it through the throat or pharynx to carry it out at the gut) and venom glands (for producing venoms to kill prey), blood circulatory systems (comprising heart and blood vessels), the central nervous system with the brain ganglia

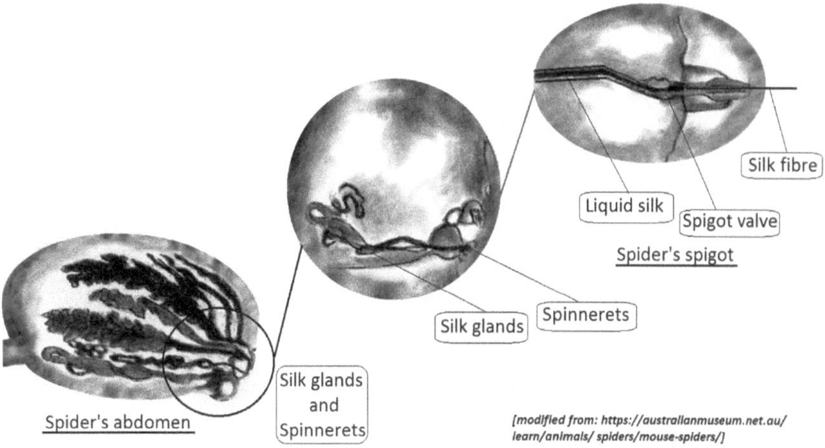

Figure 4.1 Spider abdomen, silk glands, spinnerets, and spigots.

(with numerous nervous tissues), reproductive systems, and silk glands, which are made up of 4–6 types of silk glands (viz. ampullate, flagelliform, tubuliform, aciniform, aggregate, and piriform) that are responsible for yielding protein to spin silk webs (Foelix, 2011). The diagrammatic sketch of silk glands, spinnerets, and spigots in the abdominal region of the spider and the extruding silk fiber is presented in Fig. 4.1. As a member of a community of animals with legs or arthropods, a spider's skin is made of multilayered cuticular coverings, the outermost layer of which is a very rigid structure, formed from the cross-linking of special proteins, and chitin, a long-chain polysaccharide (Davies et al., 2013). The wax coating on the outermost layer of skin of arachnids stops moisture loss from the body. From outward cuticles, like protective organs, it maintains the attachment of the muscle joints and tracheal tubes and regulates blood pressure and respiratory controls (Foelix, 2011). The inner edges of the cuticles are attached to the edges of nerves of different sense organs and integrated into the eyes and hairs and connected to the upper gut or the digestive tract extended from mouth to stomach. It also plays an important role in expediting the mating of spiders, as it acts as a sperm storage organ, or spermatheca, of spiders (Tikader, 1979; Foelix, 2011). At the time of seasonal molting, spider shed off the outermost layer of cuticular skin, which helps the inner, newly exposed, soft-layered skin become rigid to achieve outer rigidity. The skin becomes rigid slowly (arachnologists have witnessed the process carefully), and immediately after the completion of molting, spiders stop moving around until the outermost layer becomes hard. Fig. 4.1.

Scientists have also noticed that changing of blood pressure in spiders related to molting and movement, as the increased heart rate enhanced blood pressure, allowing acceleration of shedding the outermost dry and rigid scaly cuticle (Nentwig, 1987). At the same time, the differential blood pressure in the thoracic region typically flows from front to back, leading to outward limb extension. The preponderance of flexor muscles and their functional dominance over a smaller number of extensor muscles in the bloodstream-dominated situation triggers arachnid movement (Nentwig, 1987).

Spiders weave webs using silk excreted from the spinnerets that they use for multifarious activities, ranging from making homes and catching prey to rearing spiderlings (which happens occasionally) and protecting them from predatory attacks. Catching prey, immobilizing it by injecting venom, transforming the solid tissues to liquid form, and digestion are sequential, cumbersome processes for spiders (Macías-Hernández et al., 2018). The gut is the key organ in the digestive system, with two distinct parts, the foregut, which forms the stomach and its diverticula attached to the legs, and the hindgut and its diverticula, which are mainly responsible for absorption of liquid foods (Macías-Hernández et al., 2018). It is quite interesting to witness the saclike structure in the hindgut called the Malpighian tubules (functionally similar to kidneys), acting as the excretory system, and the gut ends at the anal tubercle, where guts end at the anus (Ubick et al., 2005; Denholm, 2013). The key structural difference between spiders and other insects is that insects have three primary body parts, whereas spiders have two parts, with a fused mouth and thorax called the cephalothorax (bearing six or eight simple structured eyes, mouthparts, legs, etc.) that is connected to the abdominal region by a slender waist called the pedicel that is attached by thoracic muscles. The four pairs of spider legs have two to three terminal claws (Wallace, 2017). Arachnologists have observed that spiders with two-clawed legs are mainly aggressive hunters (Ground spiders, Jumping spiders, etc.), whereas Three-claw-legged spiders are good at spinning webs (Lace-webbed spiders, Orb weavers, etc.) (Raven and Stumkat, 2003).

The spider's unique helmet-like protection, a hard cuticular, platelike protective structure in the cephalothorax, is called the "carapace." Meanwhile, the abdominal region is covered with thinner or more flexible cuticles that aid in feeding and laying eggs (Gray, 2016). The connective muscles are attached to the carapace, jaws, gut, limbs, and the skeletal structure (conspicuously nonchitinaceous in invertebrates), recognized as central lungs (Tikader, 1979). It is quite fascinating to observe the exoskeletal structure of spiders enclosing the open-blood circulatory system. Changes in blood pressure, which are normally triggered by changes in heartbeat and expansion–contraction dynamics of thoracic muscles, are regulated by this pressure-induced, flexible skeletal system known as the "hydrostatic skeleton" (Anderson and Prestwich, 1985). Arachnids deal with ever-changing issues such as variations in hunting and food intake in the seasonal cycle, changing environmental conditions such as availability of moisture, temperature, humidity, etc. This functional ability of the "hydrostatic skeleton" helps spiders manage the changing dynamics of the turgidity–rigidity equilibrium and cope with the effects of changing environmental conditions on their primitive lungs (Emerton, 1883). The important structural components and systems of spiders—web-weaving organs or spinnerets, respiratory organs (of book lungs and trachea), reproductive systems, etc.—are found in this abdominal region of the body. The structural orientation of the spider's tubular mouth is unique, having two jaws with two piercing syringe-like fangs called "chelicerae" that are involved in infusing venoms into its prey. Just beneath the spider jaws, two sets of cuticular plates forming the upper lip or "labrum" and lower lip or "labium," and a sharp platelike maxilla, which has functional and structural similarity to teeth, can be found at the front end. One pair of vital organs, the pedipalps, originally responsible for food tasting, are engaged in handling and touch sensing, as they have been functionally modified to auxiliary reproductive organs (used at the time of mating with female counterparts) (Emerton, 1883; Tikader, 1979).

The abdominal region of the spider is very important because it contains most of the critical organs that support respiratory, digestive, and reproductive systems and the lungs. Spiders breathe with book lungs comprising small openings called spiracles that become air-filled cavities that resemble the leaf-like lamellae in the pair of book lungs normally found in mygalomorphs (funnel-web spiders, trapdoor spiders, etc.) and some araneomorphs (spiders, wolf spiders, etc.). The streaming of blood (known as hemolymph for invertebrates: insects, arachnids, mollusks, annelids, etc.) through the lamellae carries out a unique process of gaseous exchange between the blood and the environment through the thin-walled skin. Tiny spiders acclimatized to wet or moist microhabitats are devoid of any breathing organs and carry out gaseous exchange through the cuticular skin of their bodies. The blood circulation happens through the open circulatory system, driven by the heart, which circulates blood throughout the body to infuse oxygen in different parts.

The reproductive organs of spiders, called testes (for males) and ovaries (for females), are responsible for yielding sperm and eggs, which are released through the gonopore or genital opening found in the genital groove present in the proximal edge of the book lungs (Emerton, 1883; Tikader, 1979). In addition to gonopore, another mating organ has been observed in female araneomorph spiders, called the epigynum. In the lower abdomen, silk glands of spiders occupy the most important functional organ, through which the liquid form of silk fluid is secreted from spigots and dried into silk threads. However, in the abdominal region, there are around 10 diverse and functionally distinct silk glands, spinnerets, and spigots (like glandular duct tubes) (Saravanan, 2006; Römer and Scheibel, 2008; Eisoldt et al., 2011) broadly categorized as follows:

- **a.** sticky, adhesive-like-catching silk,
- **b.** tangling cribellate-catching silk,
- **c.** orb-web spiral line-forming silk,
- **d.** safety line and web frame-forming silk,
- **e.** egg sac-protective silk,
- **f.** attachment disk silk, and
- **g.** swathing silk.

A diagrammatic sketch of the different silk fibers produced by female or weaving spiders is presented in Fig. 4.2.

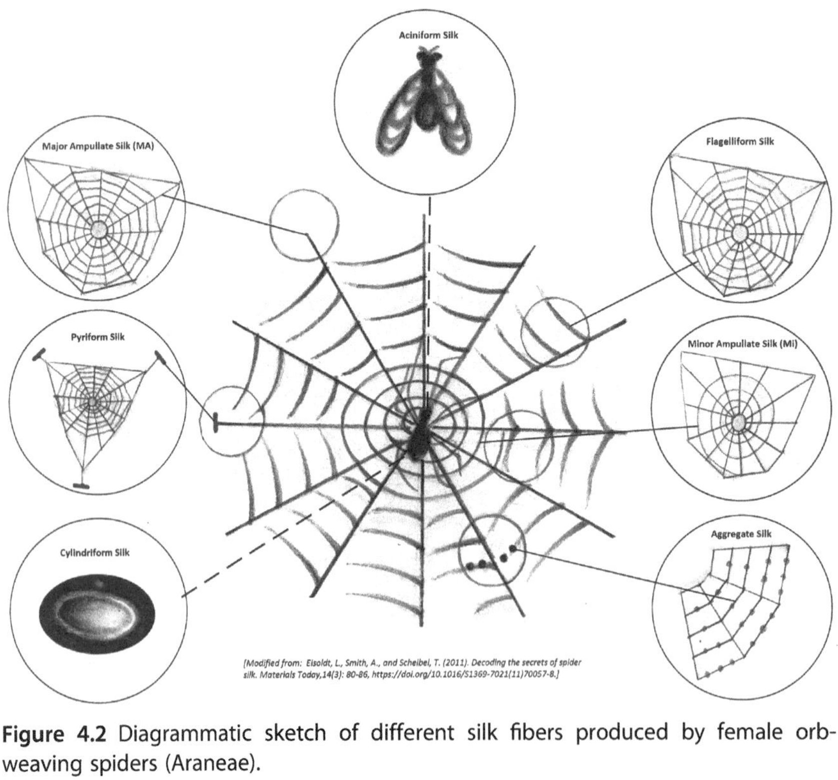

Figure 4.2 Diagrammatic sketch of different silk fibers produced by female orb-weaving spiders (Araneae).

If we look up into the trees of any woodland ecosystem, we often see spiderwebs between the branches of the same tree or two trees 2–5 feet apart. It is amazing that a tiny invertebrate creature could weave a web to cover such a vast distance, as it resembles a structural simulation of constructing a hanging bridge on a wide river resting on pillars on both sides with a wide river bed and no supporting foundation (Blackledge et al., 2011). It's a tricky method, and an amazing engineering maneuver, adopted by spiders by which they secrete fluid from the spinnerets that dries to form the silk threads used to weave spiderwebs (Spanne, 2016). Scientists have carefully observed that at the time of silk fluid secretion, the spider drags the viscous form of discharged silk fluid (which is very light), stretches it by its hind legs, and hoists its spinnerets in the stream of the blowing wind. The combination of required heat and high-velocity wind from daytime sunshine, along with the convection wind currents driven by the breeze in its microenvironment, carries the silk thread from the branch of one tree to another, and the widespread result is the formation of a wide matrix of spiderwebs (Römer and Scheibel, 2008; Ray, 2012). The morphological observation was that most spiders with two-clawed feet are not good at weaving webs, whereas those with three-clawed feet are good weavers (Tikader, 1979). Arachnologists have further witnessed that these claws in the hind legs are responsible not only for distributing the silk fluids at the time of production of silk threads but also for managing the traction of the web by controlling the grip and release of the web thread (2019). The meticulous observation of scientists on the biomechanical streaming of spider silk threads and gradual spreading of them from one branch to another shows that spider-woven silk thread is not sticky enough to stick to tree branches, and attachment is supposedly driven by static electric force (EarthSky, 2012). Arachnologists and biochemists have observed distinct parts of silk glands responsible for secreting diverse proteinaceous liquids called "spidroins" that dry into silk retrieved in the gland cavity (Garb et al., 2010). The transformation of the liquid form of protein or spidroins into silk fibers happens in two stages (Römer and Scheibel, 2008):

A. The opaque form of a proteinaceous liquid of silk sap traverses through a narrow duct, within which the inner linings of the duct absorb the water of the proteinaceous liquid and convert it into a crystal-clear fluid.

B. In the next phase, the crystal-clear fluid goes through a biochemical "acid bath" process inside the narrow duct and is subjected to upward pulling (under stress) followed by application of a coating of lipid, thus completing the change in state from liquid proteinaceous crystal-clear fluid to dry silk fibers.

Although spider-woven silk threads from the spigot duct typically have a dried fiber appearance, they will appear in liquid, sticky form after being discharged from the spigot duct when produced by orb weavers using their aggregate glands.

Biochemically, the silk fiber yield of a spider consists of heterogeneous spidroin protein; the thickness of the fibers varies from 0.001 to 0.004 nm in diameter (Römer and Scheibel, 2008). Scientists have witnessed the unique biomechanical configuration of spider-spun silk threads with a unique combination of elasticity and rigidity. When the elasticity of the silk fiber (with a stretched-coil appearance) is attributed to the presence of glycine peptides, the rigid appearance of the silk fibers is effectuated by the presence of interspersing bricklike protein bodies, alanine peptides. Most often, spiderwebs act like biochemical indicators. Scientists and engineers engaged in geological explorations and mining engineering-related explorations in Australia are primarily engaged in the qualitative assessment of water-soluble ions and the spider-inhabited sites of caves and caverns to assess the presence of metallic ions of lead and zinc content in the limestone deposits in potential mining zones (Hose et al., 2002).

Although the structural formation of the orb web is very intricate, its functional analogy is fascinating, and when it traps a fast-flying insect, it reminds us of its functional resemblance to low-impact trampolines (where the chance of bounce-back impact on the trampoline is minimal). Meticulous studies of the structural formation of orb webs put out by arachnologists have found that the orb web is biomechanically stabilized by a unique structural formation of exterior frame lines attached to the radial silk threads (resembling the spokes of the rim of a wheel) that maintains the spreading effect of the sticky lines stretching across the entire cobweb in which prey insects usually become entangled. The key silk glands (called ampullate silk glands) responsible for exuding the dragline silk (also recognized as "safety line") form the radial lines (the spokelike structures) and frame of the Orb web spiders (Boutry et al., 2011; Andersson et al., 2013). The exclusive biomechanical investigations revealed that the tensile strength of it is much higher than steel and most useful artificially formed fibers ("Kevlar" is the only exception, which has been used for manufacturing bulletproof jackets or shields). In contrast to dragline silk, the elasticity of the spiral silk-line exuded from the "flabelliform glands" is greater, but the strength of these spiral silk lines is less than that of frame-line silk (Garb et al., 2019). Thus, the combined action of frame-line silk and spiral silk-line makes the orb web strong but stretches the meshwork that can entangle the fast-moving insect and absorb the immediate thrust of the fast-moving object. If we look at the simple elucidation of the shock absorption potential of the impact of the artificial fiber Kevlar of a high-speed bullet fired from a firearm without bouncing back is a functional simulation of the landing of a fast-moving insect and getting entangled in the orb web (Garb et al., 2019). Thus, the high-impact absorbing potential and stretching ability stop the prey from rupturing the orb web while bouncing back. On the other hand, the exudation of the "sticky silk" by the aggregate silk gland, which exist in a viscous, sticky state for a while after exudation from the aggregate glands, act as the sticky trap, from which the prey failed to fly away and devoured by the spiders, ultimately (Moon, 2018).

Once the prey becomes entangled in the orb web, the web frame-line contracts to yield a primary signal to spiders with poor vision that it has prey caught in its web. The twisting and struggling efforts of the prey to free itself from the sticky web usually creates oscillating signals/waves that eventually drift and transmit through the radial lines to the spider from the location of the twitching insects to the spider as it moves toward the prey and injects its venom by inserting fangs into the prey to kill or immobilize it before devouring it (Rovner, 1980). Although the spiderweb is primarily used to trap insects, it is also used for multifarious activities. Apart from their role in hunting food, spiderwebs wrap around spiders when strong wind currents blow them to distant places and so they can land safely in new habitats to invade locations or expand areas of distribution. For safety purposes, spiders use these webs as shields to temporarily block the opening of their dwelling places. Spiderwebs have also been used to wrap up spider egg sacs. Scientists have carefully observed that if spiders find that old webs do not have functional potential, they dissolve them, consuming the raw materials, and recycled the old ones to form new ones, serving their purpose of living (Becker, 2018).

The critical observations of arachnologists have distinguished the structural and functional distinction of silk threads when spiders form the different sections of a web (Bittel, 2019). For example, it has been referred that a "Huntsman spider" use to weave a special structural pattern, called attachment disk (Wolff et al., 2017), whereas another orb-weaver formed a pattern of weaving the smaller length of silk threads by the Black-yellow garden spiders, secreted by a pair of spinnerets, located at the anterior-lateral location of the frontal end (Hadley, 2019). Those anterior-lateral spinnerets end up with piriform spigots. The biomechanical studies of the web ascertained that the major ampullate gland, adjacent to piriform spigots, secreting silk fluid, dried to form the strongest part of the web, called dragline or safety line, adhered to the attachment disk. The fusing points of the dragline to the attachment disk act as the construction anchor to maintain the structural integrity of the web. The spider uses this safety line as a walking trail when it moves from one end to the other end of the web to catch the prey. The careful observation of arachnologists further revealed that the silk threads secreted from the aciniform spigots use the wrapping-up technique to entrap prey in hunting or egg-wrapping (forming cocoons) activities during the reproductive cycle (Becker, 2018).

As the spider builds most parts of the web surface with the insect-trapping potential of the silk lines, the spider could become trapped in the sticky part of its own web. But that never happens. Careful observation of the resting posture and state of locomotion elucidated further that the tip of the jointed leg (specifically the middle claw and its toothed bristle), rather than the rest of the body, contacts the silk lines of the web (Tikader, 1979). It's quite amazing to watch the cleaning of the leg tip by the jaws to leave it clean and free from the sticky fluid of the catching silk, so it never becomes stuck in the orb web. However, arachnologists further promulgate two distinct ways that spiders have evolved catching prey with the help of their webs. According to the structural formation of the spider's web, it has broadly categorized into three categories (Hawkes, 2015; Lawania and Mathur, 2015; Bittel, 2019):

1. Orb web: As the orb weavers (e.g., garden spider) are poor in eyesight, they spun the web with the help of touch-mediated sense organs. 1980, George W. Dunne and Roland F. Eisenbeis, described in the *Nature Bulletin* (volume 165) "Beneath her abdomen are six spinnerets that can be extended or compressed and used like the fingers of a human hand. Each spinneret has 'faucets' and 'spools' connected by tiny tubes to several types of glands." The spinnerets exuded fluid form of the silk, dried in air and forming the orb-weaver foundation by radial and framework thread, formed by dragline silk. Then the orb weavers smeared the dots on dragline thread with the viscous, sticky silk to make the orb web ready to catch prey. Typically, it takes an hour to an hour and a half to finish weaving the Orb web spiders.
2. Sheet and funnel webs: In the same edition of *Nature Bulletin* (volume 165), George W. Dunne and Roland F. Eisenbeis (1980) further state, "Each sheet extends outward from a funnel-like opening beneath which the spider lurks in waiting for grasshoppers and other insects that may alight upon it. She constantly enlarges the web, and if undisturbed, it may

become a square yard in the area." Arachnologists have observed that spiders (e.g., grass spider) in the woodlands and forest vegetation areas spun the horizontally formed funnel web underneath its dwelling place in the cracked fissure of rock crevices or thick vegetation areas, where the funnel ranges from 10 to 12 inches long, and the horizontal sheet on its top ranges from 1.5 to 3 feet in diameter.

3. Irregular web: It is a common feature that we are usually familiar with house spiders, also known as comb-footed spiders for their special-shaped feet with curved bristles that are used to toss the silk threads of the web to entrap flying insects. Comb-footed spiders, normally found in sheds, barns, porches, warehouses, cellars, etc., are involved in forming the irregular shapes of the webs and are competent enough to catch flies, mosquitoes, and other insects in their webs to control their populations biologically. House spiders can even kill small-sized mammals such as mice to suck body fluids. In the same edition of *Nature Bulletin* (volume 165), George W. Dunne and Roland F. Eisenbeis (1980) further stated, "*A typical web, usually built at night, has a central sheet of densely woven silk which serves as a hiding place and is anchored by numerous guy lines that are long and strong.*"

It has been hypothesized that around 10 million years ago, the evolution of orb-web spiders used the ampullate silk to form different funnels, sheets, tangles, etc. From their secondary level of reviews and simulations of the web thus constructed, arachnologists have contemplated that the maiden web was horizontally arrayed and was certainly not the orb web, and the orb web is construed to have evolved only once. However, according to state-of-the-art catching of distinct silk threads in the web formation (that act as a trap), silk is categorized as (1) cribellate or woolly silk or (2) ecribellate or sticky silk (Coddington and Levi, 1991; Vollrath and Selden, 2007; Wessel, 2018).

a. Cribellate silk: The widow spiders and the common house spiders are net weavers of the shawl and sheet webs, with the help of cribellate silk fibrils. Several small silk glands beneath the spinning organs of the spider are recognized as cribellum, and the silk fibrils or fine threads produced by these cribellum called cribellate silk. The diameter of the cribellate silk is around 0.00001 mm. The cribellum is evolved from anterior-median spinnerets, and the surface of the cribellum is covered with thousands of spigots yielding fine-threaded silk fibrils, and those numerous fibrils fuse to form a single thread of cribellum. Cribellate spiders possess a set of toothed bristles, called the calamistrum, at the metatarsal segment of the second to last and last leg, and the calamistrum spreads and combs out silk fibrils to form a cribellate silk thread that is initially secreted from spinnerets and cribellum in liquid form (Kullmann, 1972). The evolutionary biologists observed that spiders belonging to the superfamily—Araneoidea had been evolved as cribellate but lost cribellum in derived lineages in the course of evolutionary changes (e.g., Jumping spiders, Wolf spiders, Huntsman spider, etc.). The evolutionary and molecular biologists have hypothesized that around 136 million years ago, the evolutionary combination of flagelliform silk threads to the aggregate silk protein glue, leads to the evolutionary surge of spider species belonging to Araneoidea.

b. Ecribellate silk or sticky: From the perspective of evolutionary biology, if the evolution of cribellate silk formation considered the ancestral traits of specialization of formation of the web for catching the, evolved among the ancestral lineages of the current descendants, the evolution of the webs from sticky silk threads, formed by the derived lineage of spiders, considered the recent evolutionary innovation of spiders. Arachnologists carefully observed that the newly acquired innovation of web-spinning techniques using sticky silk is a unique method of web formation that is normally found in the ancestors of Comb-footed and Orb-weaver groups. The most interesting feature of the Ecribellate web weavers that the liquid secreted from spinnerets of these spiders remained in the viscous, sticky state for a long time in normal temperature rather than transformed into dried fibrous or woolly silk threads. The critical observation of its secretion from the spinnerets revealed that the fluid of sticky silk had been exuded from the "triad" form of spigots, connected with the posterior—lateral spinnerets. This critical observation of scientists on Orb-weaving spiders further revealed that from the linearly formed three spigots or the triad (which is consist of the central spigot flanked by two lateral spigots either), the central spigot forms the fibrous supporting line of the silk, fluid secreted by the flagelliform silk gland, while the two lateral spigots made a coat liquid silk on the fibrous line, secreted from two aggregate silk glands (Blamires et al., 2014). In Cobweb spiders, the supporting line is formed from the fluid secreted by ampullate glands (Blamires et al., 2014). However, the entire process is responsible for forming a double sticky line containing glycoprotein. Biochemists noticed that the sticky silk captures moisture from the surrounding atmosphere. Scientists have promulgated this hygroscopic potential of sticky silk or intermittent wetting feature of the spiral line is another autostretching evolutionary innovation of the orb weavers that help the contemporary lineages of the ancestral spider groups be recognized as an evolutionarily advanced group of spiders, with a better and diverse range of prey-catching abilities that made the and widespread dispersal in a wide range of habitats.

However, it is quite amazing for arachnologists and molecular biologists to understand the mystery behind this. They wondered how an invertebrate creature with poor or no vision, like a spider, could successfully weave an abstract weblike structure in space using its primitive (touch-induced sense potential) sense organs. To unlock the molecular footprints of the web-spinning process of spiders, scientist Andrew Gordus in the Department of Biology at John Hopkins University, USA, set up surveillance in his experimental laboratory, where the entire web-making process and physiological, biobehavioral activities have been monitored and recorded using an infrared camera (Hirsch, 2017). The primary objective of this kind of investigation done by Dr. Gordus helped achieve the key information about a tiny invertebrate creature (as an experimental model, Gordus has chosen a spider species *Uloborus diversus*) to discover its web-making process with its primitive, small brain size and preliminary genetic profile (Hirsch, 2017). When a reptile, bird, or small mammal makes its nest, it inevitably uses its inheritable "spatial memories," an extant inheritance from its predecessor that has been vertically transcended generation after generation in its evolution (Sotelo et al., 2016). For other animals, mostly vertebrates (birds, reptiles, amphibians, etc.), it is much easier to build a nest site either by accumulating external materials (bird's nest) or removing existing materials (mud burrows) on a supporting base, either terrestrial, aquatic, or arboreal. But if we look at the web-building process of spiders, we witness a unique combination of using the spatial memory of its tiny brain, maximal use of physiological and biobehavioral potential as a resource to build an abstract form of structure like a web in the air that would help it survive and protect it from predators as well as complete its reproductive cycle and migrate to distant locations. By referring an identical brain profile of a *Drosophila melanogaster* (fruit fly), which has 100,000 brain neurons, Dr. Gordus simply wondered: "How does a brain encode all these behaviors in the correct sequence?" (Hirsch, 2017). To track the behavioral action of spinning the web, Dr. Gordus, and his fellow researcher, Nick Wilkerson developed a computer program that would monitor the entire action of the developmental process of web-weaving efforts concerning its legs and other physical and physiological movements. In order to findout the molecular switch of web-spinning gene, the gene sequencing process of the spider genome is an essential precondition and Dr. Gordus and his associates are engaged in that endeavor, currently. The main objective of the scientists in Dr. Gordus's laboratory in John Hopkins University is to establish the link between behavioral changes during spiderweb-building process in relation to genome in order to find out the specific molecular factor, encoded by the spider's DNA, in relation to its web-building process (Hirsch, 2017). Dr. Gordus and his associates have a grander dream—to develop and implant a molecular probe in the tiny brain of the spider to track down the nerve impulses generated by its brain to direct the physiological course of action of arachnids that allowed them to spin abstract webs in the air for a long period ranging from the inception of its evolution some 400 million years ago (Bittel, 2019). It's true that except for Halloween (when we decorate our homes with fake spiderwebs), we don't feel comfortable seeing spiderwebs in our dwelling places, but the spider's mystic and amazingly spun and bioengineered web should make us curious to unlock and unfurl the molecular footprint of the origin and evolution of spiderwebs, as well as spiders, encrypted in its genome.

References

Anderson, J.F., Prestwich, K.N., 1985. The physiology of exercise at and above maximal aerobic capacity in a theraphosid (tarantula) spider, Brachypelma smithi (F.O. Pickard-Cambridge). J. Comp. Physiol. B 155, 529. https://doi.org/10.1007/BF00694442.

Altherr, R., 2019. The Legs of Spiders Move Using a Combination of the Hydraulic Pressure of Body Fluid and Muscle Flex. https://asknature.org/strategy/leg-uses-hydraulics-and-muscle-flex/.

Andersson, M., Holm, L., Ridderstråle, Y., Johansson, J., Rising, A., 2013. Morphology and composition of the spider major ampullate gland and dragline silk. Biomacromolecules 14 (8), 2945–2952. https://doi.org/10.1021/bm400898t.

Barth, F.G., 2002. Spider senses – technical perfection and biology. Zoology 105 (4), 271–285.

Becker, A., 2018. How to Identify Spider Egg Sacs. https://sciencing.com/identify-spider-egg-sacs-4886667.html.

Bittel, J., 2019. How Spider Silk Is One of the Most Versatile Materials on Earth. www.nationalgeographic.co.uk/animals/2019/09/how-spider-silk-one-most-versatile-materials-earth.

Blackledge, T., Kuntner, M., Agnarsson, I., 2011. The form and function of spider orb webs. Adv. Insect Physiol. 41, 175–262. https://doi.org/10.1016/B978-0-12-415919-8.00004-5.

Blamires, S., Sahni, V., Dhinojwala, A., Blackledge, T., Tso, I.-M., 2014. Nutrient deprivation induces property variations in spider gluey silk. PLoS One 9, e88487. https://doi.org/10.1371/journal.pone.0088487.

Boutry, C., Řezáč, M., Blackledge, T.A., 2011. Plasticity in major ampullate silk production in relation to spider phylogeny and ecology. PLoS One 6 (7), e22467. https://doi.org/10.1371/journal.pone.0022467.

Coddington, J.A., Levi, H.W., 1991. Systematics and evolution of spiders (araneae). Annu. Rev. Ecol. Systemat. 22, 565–592.
Davies, G .J., Knight, D.P., Vollrath, F., 2013. Chitin in the silk gland ducts of the spider Nephila edulis and the silkworm Bombyx mori. PLoS One 8 (8), e73225. https://doi.org/10.1371/journal.pone.0073225.
Denholm, B., 2013. Shaping up for action: the path to physiological maturation in the renal tubules of Drosophila. Organogenesis 9, 40–54. https://doi.org/10.4161/org.24107.
Dunne, G.W., Eisenbeis, R.F., 1980. Spider Webs. Nature Bulletin No. 165. Forest Preserve District of Cook Country. http://www.newton.dep.anl.gov/natbltn/100-199/nb165.htm.
EarthSky in Earth, 2012. How Does a Spider Spin a Web between Two Trees? https://earthsky.org/earth/how-does-a-spider-spin-a- web-between-two-trees.
Eisoldt, L., Smith, A., Scheibel, T., 2011. Decoding the secrets of spider silk. Mater. Today 14 (3), 80–86. https://doi.org/10.1016/S1369-7021(11)70057-8.
Emerton, J.H., 1883. The Structure and Habits of Spiders. Sagwan Press, p. 124 [Revised Edn. 2018].
Fabian-Fine, R., Hoger, U., Seyfarth, E.-A., Meinertzhagen, I.A., 1999. Peripheral synapses at identified mechanosensory neurons in spiders: three-dimensional reconstruction and GABA immunocytochemistry. J. Neurosci. 19 (1), 298–310. https://doi.org/10.1523/JNEUROSCI.19-01-00298.1999.
Foelix, R., 2011. Biology of Spiders. Oxford University Press, p. 432.
Garb, J., Ayoub, N., Hayashi, C., 2010. Untangling spider silk evolution with spidroin terminal domains. BMC Evolut. Biol. 10, 243. https://doi.org/10.1186/1471-2148-10-243.
Garb, J.E., Haney, R.A., Schwager, E.E., Gregoric, M., Kuntner, M., Agnarsson, I., Blackledge, T.A., 2019. The transcriptome of Darwin's bark spider silk glands predicts proteins contributing to dragline silk toughness. Commun. Biol. 2, 275. https://doi.org/10.1038/s42003-019-0496-1.
Gray, R., 2016. There Are Spiders that Encase Their Bodies in Thick Armour. http://www.bbc.com/earth/story/20161102-there-are-spiders-that- encase-their-bodies-in-thick-armour.
Hadley, D., 2019. Spider Silk Is Nature's Miracle Fiber. https://www.thoughtco.com/what-is-spider-silk-1968558.
Hawkes, A., 2015. Learn How to Recognize a Spider by its Web:Your Guide to Nature's Most Perfect Traps. https://baynature.org/article/spiders/.
Hirsch, A., 2017. Watch a Spider Use its Tiny Brain to Build a Web. https://www.futurity.org/spider-webs-brains-1635792/.
Hose, G.C., James, J.M., Gray, M.R., 2002. Spider webs as environmental indicators. Environ. Pollut. 120 (3), 725–733.
Lawania, K.K., Mathur, P., 2015. Study on the pattern and archetecture of spider's web with special reference to seasonal abundance in eastern region of Rajasthan, India. IOSR J. Environ. Sci. Toxicol. Food Technol. 9 (1), 01–09. e-ISSN: 2319-2402, p-ISSN: 2319-2399. www.iosrjournals.org.
Kullmann, E.J., 1972. The convergent development of orb-webs in cribellate and ecribellate spiders. Am. Zool. 12, 395–405.
Macías-Hernández, N., Althey, K., Tonzo, V., Wangensteen, O.S., Amedo, M., Harwood, J.D., 2018. Molecular gut content analysis of different spider body parts. PLoS One 13 (5), e0196589. https://doi.org/10.1371/journal.pone.0196589.
Moon, M.J., 2018. Fine structure of the aggregate silk nodules in the orb-web spider Nephila clavata. Anim. Cell Syst. 22 (6), 421–428. https://doi.org/10.1080/19768354.2018.1546227.
Nentwig, W. (Ed.), 1987. Ecophysiology of Spiders. Springer Verlag, Berlin Heidelberg, p. 448.

Pomeroy, R., 2013. Spiders, and Their Amazing Hydraulic Legs and Genitalia. https://www.realclearscience.com/blog/2013/02/spiders-their-amazing-hydraulic-legs-and-genitals.html.

Raven, R.J., Stumkat, K., 2003. Problem solving in the spider families miturgidae, ctenidae and psechridae (araneae) in Australia and New Zealand. J. Arachnol. 31, 105–121.

Ray, C.C., 2012. Blowing in the Wind. https://elsevierbooks.proofcentral.com/en-us/index.html?token=.

Römer, L., Scheibel, T., 2008. The elaborate structure of spider silk: structure and function of a natural high performance fiber. Prion 2 (4), 154–161. https://doi.org/10.4161/pri.2.4.7490.

Rovner, J., 1980. Morphological and ethological adaptations for prey capture in wolf spiders (araneae, lycosidae). J. Arachnol. 8 (3), 201–215. www.jstor.org/stable/3705002.

Saravanan, D., 2006. Spider silk-structure, properties and spinning. J. Tex. App. Tech. Manag. 5, 1–20.

Spanne, A., 2016. 8 Silkily Engineered Facts about Spider Webs. https://www.mentalfloss.com/article/87870/8-silkily-engineered-facts-about-spider-webs.

Sotelo, M.I., Daneri, M.F., Bingman, V.P., Muzio, R.N., 2016. Telencephalic neuronal activation associated with spatial memory in the terrestrial toad *Rhinella arenarum:* participation of the medial pallium during navigation by geometry. Brain Behav. Evol. 2016 (88), 149–160. https://doi.org/10.1159/000447441.

Tikader, B.K., 1979. Handbook Indian Spiders. Zoological Survey of India, Calcutta, p. 251.

Ubick, D., Paquin, P., Cushing, P., Roth, V., 2005. Spiders of North America: An Identification Manual. American Arachnological Society, p. 377. www.researchgate.net/publication/260336753_Spiders_of_North_America _An_Identification_Manual.

Vollrath, F., Selden, P., 2007. The role of behavior in the evolution of spiders, silks and webs. Annu. Rev. Ecol. Evol. Syst. 38, 819–846. https://doi.org/10.1146/annurev.ecolsys.37.091305.110221.

Wallace, T., 2017. The Science of Spider Paws. https://cosmosmagazine.com/biology/the-science-of-spider-paws/.

Walcott, C., 1969. A spider's vibration receptor: its anatomy and physiology. Am. Zool. 9 (1), 133–144. https://doi.org/10.1093/icb/9.1.133.

Walter, A., Bechsgaard, J., Scavenius, C., Dyrlund, T.S., Sanggaard, K.W., Enghild, J.J., Bilde, T., 2017. Characterisation of protein families in spider digestive fluids and their role in extra-oral digestion. BMC Genom. 18 (1), 600. https://doi.org/10.1186/s12864-017-3987-9.

Wessel, L., 2018. Sticky Science: The Evolution of Spider Webs. https://www.scientificamerican.com/article/sticky-science-the-evolution- of-spider-webs/.

Wolff, J.O., Herberstein, M.E., 2017. Three-dimensional printing spiders: back-and-forth glue application yields silk anchorages with high pull-off resistance under varying loading situations. J. R. Soc. Interface 14, 20160783.

CHAPTER FIVE

The tangled equilibrium of predator and prey: The hunter becomes the hunted

Abstract

The ecological dynamics of the predator—prey relationship in arachnids have often been seen as a reversible act of hunting that scientists recognize as a swapping ball game. The swapping of positions by predators and prey in this dangerous hunting game often results in altered trophic dynamics. Altered energy flows have been observed to reverse direction from one end to another and vice versa. The evolution of counteradaptive defense mechanisms against predators among prey communities has helped them ensure their immediate survival. The acquired traits are passed on to the next and following generations, resulting in their evolution into a derived group of predators that prevail over their erstwhile predators to make them a derived group of prey over time.

Keywords: Arachnids; Ecological dynamics; Predator—prey relation.

Predator—prey arms races are often asymmetrical, as a prey organism is under stronger selection pressure (Brodie and Brodie, 1999). As a result, prey possess various primary and secondary defences (Edmunds, 1974) that make them hard to capture or even dangerous for a predator. Hunting of such prey may be too costly for a predator, as it faces a risk of injury or even death.

Ondřej Michálek, Yael Lubin & Stano Pekár

In the course of autecological investigations, specifically, the ecological dynamics of the predator—prey relationship of large spider species, such as *Poecilochirus senilis* in the Negev desert, Israel, arachnologists and evolutionary biologists have noticed the wide range of trophic diversity and how *P. senilis* has evolved to eat a diverse group of insects and gradually adapted to hunting a small range of spiders (Michálek et al., 2019). The dynamics of the predator—prey relationship depend on the act of hunting, and it is a swapping ball game, as predators often swap their positions with prey in this dangerous game of hunting. For a long time, this action and counteraction dynamics-oriented predator—prey relationship caused scientists to contrive a random process to explain it, but critical observations of

Spider Evolution
ISBN: 978-0-323-90041-6
https://doi.org/10.1016/B978-0-323-90041-6.00005-X

predator–prey dynamics from an evolutionary perspective cleared the way to acknowledge that changing the status of the hunter to hunted is a behavioral adaptive evolutionary process that animals have gone through over time (Durrett and Mayberry, 2010).

In support of their contemporary findings, arachnologists have stated that *"specialized araneophagic (spider eating) predators have evolved intriguing hunting strategies that allow them to invade spiders' webs by adopting a stealthy approach or using aggressive mimicry"* (Michálek et al., 2019). Thus, contemporary experimental analysis by arachnologists on the biobehavioral evolutionary adaptation of *P. senilis* has helped them recognize this obligate insectivorous member of Araneae as facultative and araneophagic (hunts small-sized spiders and eats them), and scientists have contemplated that *P. senilis* might have evolved as a typical insectivorous spider with an evolutionary pattern similar to that of other spiders, but over time, this particular species may have drifted from the typical pattern of interspecific struggle and tried to establish its supremacy over others to get into an intraspecific struggle of existence and use its higher-potential venom to immobilize its intraspecific competitors, along with a new learned skill—how to use the "forelegs," "mouthparts," and "swarthy glue silk" to hunt small-sized spiders and eat them.

It seems that spiders have evolved with a series of defense safeguards—biobehavioral, physical, physical, and biochemical—that make them nearly impossible for potential predators to hunt, but the matter of fact is that if a group of animals has evolved with special adaptive defense mechanisms to protect them from predatory attacks, nature turns its metaphoric wheel around over time, which helps evolve potential predators with developed counteradaptive defense strategies, and those kinds of newly acquired traits are generally passed on from one generation to the next to render the derived group of animals with new traits to conquer invincible prey. An interesting observation in this experimental venture was that juvenile members of *P. senilis* have evolved with an araneophagic trait that allows them to hunt larger-sized spiders (maybe more powerful in terms of physical strength), subdue them by injecting venom, and ultimately eat them (Michálek et al., 2019).

Another reality is that the evolutionary arms race is part of the struggle of existence that attributes the supremacy of a particular species over another and often helps establish the supremacy of one geographical race of a particular species over another geographical race of an adjacent or distant location of the same species. For example, Caribbean athletes, specifically athletes of Jamaican origin, are some of the fastest athletes in the history of the track and field world championships in short-distance running events (such as the 100 m sprint).

Spiders have many predators, so they adopt many defense strategies to save themselves from attack. Most spiders are slow-moving species with respect to other invertebrates, as they don't have leg muscles, so in the stiff challenge of predatory attacks, they cannot move fast enough to escape the reach of potential predators. To offset this downside, spiders engage hydraulic systems that enable them to jump and cross long distances quickly (Altherr, 2019). To protect themselves, most spider species rely on their primary defense strategy, which evolved as camouflage through unique polymorphic colorations to help them blend in with their microenvironment so predators can't find them. This defense mechanism also helps them take on a deadly look to scare their predators (Geyer, 1967). These spectacular external polymorphic colorations, with different patterns among the same species, not only confuse potential predators of spiders but also challenge arachnologists in applying their taxonomic judgment.

Many arachnid members have evolved with biochemical fluid that makes their tissues distasteful. After predators of spiders have attacked and killed them, they are unable to eat them because of their distasteful tissues. After such an attack, the predators immediately leave the carcass of the dead spider, and that negative image is stored in their memory so that the next time, the potential predators will avoid that kind of distasteful arachnid prey (distasteful mimicry) during their hunting operations (Stevens, 2010). Spiders have fine-bristle hairs on their outer body surface, and some but not all of those hairs act like filiform spines. To escape the grip of predators, spiders throw these spinous hairs to irritate predators to lose their grasp, which allows spiders to flee from their predators (Agnarsson et al., 2009).

The spiderweb is an abstract form of art that a spider spins with its silk that helps it be recognized as a predator entrapping prey to meet its feeding demand to survive and at the same time is used as a defense shield when it becomes the prey of some other species' predatory attacks (Yong, 2017). When a spider senses the existence of a potential predator, it receives sound waves through its vibration-mediated sense reception systems, then immediately takes refuge in the safety hub of the web and waits there until the predator has left its place nearby (Edwards, 2011). On the other hand, spiders smear their silk threads with an ant-repellent chemical, called 2-pyrrolidinone, to deter ants from invading their webs, as ants are aggressive predatory insects that would easily invade spiderwebs in the absence of spiders using this biochemical as an ant biorepellent (Zhang et al., 2011).

The highest level of the inbuilt defense mechanism of a spider depends on the nature of the toxicity of its venom, a type of special protein physiologically produced in the venom glands of its abdomen (Cardoso and Lewis, 2019). If the spider fails to avoid a predator attack, it uses its fangs (analogous to injection needles) present in the chelicera or the mouthpart of its cephalothorax to inject toxic venom to immobilize or kill the predator. Apart from these basic defense measures, spiders have gradually evolved special features such as hiding themselves from exposure to predators in daytime by taking refuge behind rocks and shelter inside the little cracks of rock crevices or holes in trees and remaining there in a state of suspended animation. Many spiders have evolved with less efficient defense systems internally and the spiked type of armor on the body exterior that saves them from being easily immobilized and devoured by predatory birds and other small reptilian and mammalian predators (Roach, 2001). Thus, spiders took a long walk on the evolutionary trail to stay on the side of predators as a survival edge over their prey rather than taking a position among prey, and this has helped them occupy nearly all diverse ecological niche categories dispersed over terrestrial habitats worldwide.

We human beings prioritize airborne sounds as the primary mode of communication conceived within the animal kingdom, with the origin and evolution of a higher group of plants, advanced in nature, that originated and evolved more recently than the primitive animals, the predecessors of the advanced group of animals that evolved on Earth in prehistoric time and depend on substrate-borne signaling as the primary mode of communication (Virant-Doberlet et al., 2019). If we attempt to communicate with someone, we just call him or her by making an audible airborne noise to send our message. But we could also detrmine the existence of a particular person or animal by deciphering the nature of its specific vibrations from the floor or substrate when a particular person or animal strides on it following a particular walking style. However, scientists have noticed that in maintaining the predator—prey relationship, such substrate-borne signaling plays a primary mode of interspecific communication. In support of this view, Dr. Meta Virant Doberlet, Head of the Department of Organisms and Ecosystems Research, National Institute of Biology, Ljubljana, Slovenia, referring to her recent research observations, said, *"Eavesdropping and exploitation of vibrational signals used in sexual communication have been so far largely neglected; however, existing studies show that generalist arthropod predators can intercept such signals emitted by insects to obtain information about prey availability and use that information when making foraging decisions."* (Virant-Doberlet et al., 2019). Scientists have observed from field-based experimental observations that in the foraging process of the predator, the vibrational signals from a substrate are gender-biased, as the males make vibrations of higher amplitude that last longer than those of females and further helps the potential predator get closer to the prey and also makes males more vulnerable than their female counterparts (Virant-Doberlet et al., 2019).

However, to elucidate the relationship of the predator—prey dynamics of spiders, ecologists have observed that several spider species have come closer to bodies of water in search of a diverse group of insects that they hunt. Praying mantises, aka long-necked assassin bugs, are a big group of arthropods (around 2400 species) under the class Insecta that are predominant in all forest ecosystems, and they feed many spiders and manage their subsistence (Yong, 2016). On the other hand, many amphibians, such as frogs and toads, and many reptiles, such as salamanders and lizards, gather in the same habitat in search of prey, mainly insects along with many spiders to hunt. So in the search for food, small-sized spiders are recognized as hunters and are hunted often. In tropical forest and coastal ecosystems, reptiles such as snakes, monitor lizards, wild birds, and seabirds and small primates such as bats, otters, and monkeys manage their subsistence using a diverse source of tropic levels, so along with plant-based food-resources like fruits, the representatives of reptiles, avifauna, and primates hunt several spider species along with other insects as part of their daily meal (von May et al., 2019). The potential hunting ground for spiders, the moist habitats for prevalent distribution of a bigger crown of insects, is also a congenial abode of fungi and molds that often cause an infestation of fungal disease to spiders and biologically limit their population growth. Many wasps and ticks act on spiders to interfere with normal living off their host spiders. In predator—prey dynamics, intraspecific competition of spiders has been given a new dimension as spiders have become araneophagic and involved in hunting, which is considered a behavioral adaptation of cannibalism. Another mode of gender-related practice of cannibalism has been observed among spiders related to gender-identity and dimorphism. Scientists observed that larger-sized female spiders usually devour the male members at the end of the mating process, and this has been assumed to be an integral part of successfully completing the reproductive cycle (Britt, 2008).

In the current regime of Anthropocene, anthropogenic activities are beings that are considered one of the highest-level predatory challenges that spiders have survived through the ages. The gradually increasing use of toxic insecticides and pesticides will inevitably diminish the size of the populations of insects and agricultural pests to save crop yields but diminish the availability of insects as prey to the spiders that hunt them. It should be mentioned here that in some parts of Southeast Asia, such as Cambodia, Thailand, frightful spider species like Tarantulas have been cooked like a deep-fried delicacy and have a commercial craving in the regional food market (Erict_Culinarylore, 2016). The shrinkage of a wide range of food resources for spiders due to anthropogenic actions not only jeopardizes the existence of the spider population but also forces them to explore alternative sources of energy to survive. Therefore, arachnologists have tried to determine whether the evolutionary surge of araneophagic food habits among a large number of spider species is simply a matter of chance or a matter of evolutionary advantage over the conventional insectivorous food habits of most spiders (Earthkind, 2017). Apart from the indirect pressure of anthropogenic interference, the clearing of forest habitat or wild habitat is an enhanced spree of housekeeping affected by the web-spinning process that usually affects the catching of prey and feeding to stay alive. It has been factiously reported by arachnologists that people feed eight spiders in a year during their sleep. It sounds like a matter of sarcasm, but Rod Crawford, the arachnid curator of the Burke Museum of Natural History and Culture in Seattle, USA, says that *"Once or twice a year, someone tells me they once recovered a spider leg in their mouth."* (Sneed, 2014). Then Crawford elucidated that swallowing eight spiders in a year by sleeping people has no scientific basis and is better considered a myth, as spiders supposedly find "sleeping humans terrifying" because of of heartbeat vibrations and snoring. Crawford comments further: "Vibrations are a big slice of spiders' sensory universe" (Sneed, 2014). Qualitative experimental observation ascertained that the composite vibrations of heartbeats and snoring stalled house spiders from weaving their webs, which is an essential process of entrapping prey to eat.

From a sociobiological viewpoint, the predator—prey relationship is a simple cat and mouse chase. In their contemporary investigations on predator—prey interactions from the perspective of behavioral ecology, arachnologists stress a key factor called "encounter probability." Scientists consider three cardinal issues that enhance the thrill of the cat and mouse game by maximizing the "encounter probability," as follows (Craig and Freeman, 1991):

a. Predators need to identify the arena of prey that are reachable to hunt, or
b. Predators have the full potential to lure them (prey) so that the prey gathers close to their vicinity, and they can easily grab them, or
c. Prey must have predator-recognizing ability so they can easily flee from the ambiance of predators.

To examine the cardinal role of "encounter probability" in maintaining the equilibrium of predator–prey dynamics, Craig and Freeman (1991) coordinated an experimental set of foraging experiments of a predatory web-weaving spider, where three variables were selected for determining the effect of predatory attraction of the prey toward the predator to enter and sit on the orb-web, to recognize the orb-web, or avoid getting closer to the orb-web by these variables:

1. The random presence of sticky silk droplets in the spider orb-web
2. Brightness (in the form of reflectance) of silk threads that attracts prey to visit the web, and
3. The visibility of the orb-web pattern that attracts the prey to get closer and sit on it.

The experiment yielded a unique result, that the true identity of sticky droplets on the orb-web seems to be convincible to the prey from a close distance, but the virtual identity of the sticky silk droplets from afar seems to be like a string of pearls that lures the prey to get closer to the orb-web, and hence, it enhanced the prey capture probability (Craig and Freeman, 1991; Li, 2005). Though scientists find a close correlation between the sticky silk droplets and the brightness of the silk strings of the orb-web, the relationship among variables like the size of the droplet, size of the spider, and visual environment in which the orb-web was spun seems to be complicated. It has been noticed that in visible light exposure, the size of the droplet of sticky silk correlates with spider size. In the low light exposition, the shining of the silk thread (reflectance from silk thread) and the size of the spider have no strong correlation. The most intriguing result from this experiment is that the foraging spider spins a greater quantity of its more visible silk during nighttime than during the day (Craig and Freeman, 1991). The result from such an experimental model is that the complex exposition of light, along with visual and foraging alterations, plays a key role in the evolution of spiders and a key factor in predator–prey dynamics in a microenvironment.

In another experiment, on *Nephila clavipes*, a foraging orb-weaving spider that is abundantly distributed over tropical, subtropical, and ecotone habitats situated between the tropics and subtropics, altered the color reflectance of the (Craig et al., 1996) orb-web it spun under the exposition of different grades of color irradiance. Scientists have observed that the diverse grade of exposition of light lets predatory spiders spin orb-webs with distinct color irradiance. The critical observation of arachnologists was that under bright light exposition, *N. clavipes* weave a web with spider-silk that, with yellow pigment, helps the orb-web absorb a certain light spectrum of 500 nm wavelength, and the orb-web appears golden (Craig et al., 1996). It has further been observed by scientists that the bright yellow color acts as a visual stimulus of the orb-web and lures big crowds of herbivores and pollinating insects to sit on the orb-web and become stuck, and that leads to high foraging pressure of orb-weaving spiders on their prey. On the other hand, scientists have observed that in low light exposition, the silk thread of the orb-web can absorb a light spectrum between 400 and 500 nm. Under the low exposition of light, like the 400–500 nm wavelength of light spectra, the orb-web appears whitish, which does not attract enough insects (Craig et al., 1996). In the experimental trial, scientists used *Trigona flaviventris*, stingless bees, as prey for the potential predator *N. clavipes,* and it was observed that *T. flaviventris* was attracted by the bright yellow colored silk threads of the orb web. Because of the gathering of a big crowd of prey on the orb-web under the bright golden-yellow light reflectance, the foraging rate of *N. clavipes* was enhanced (Craig et al., 1996).

Hence, the experimental results show that *N. clavipes* has evolved with a higher rate of foraging under high exposition of the light spectrum as well as distinct types of color cues that lure prey insects to rest on the orb-web in higher numbers that lead to the foraging rate of this orb-web spider, and a group of scientists considers it to have complied with the "multiple prey hypothesis" (Ximenes and Gawryszewski, 2019). Scientists have further contemplated that the diverse group of insects are often lured by the orb-web spiders when spiders mimic colorful flowers and attract a big crowd of insects, and this kind of crowd-pulling tactic has been explained under the "flower mimicry hypothesis" (Ximenes and Gawryszewski, 2019). It has also been observed by arachnologists that changing the external colors of the same spider species creates several polymorphic species, and these polymorphic predators often attract many prey insects, which further enhances the level of foraging and is considered the "multiple predator hypothesis" (Ximenes and Gawryszewski, 2019).

When it comes to defining the spider as a strategic predator, coordinating its actions to capture the fast-moving and physically stronger insects, it comes to the formation of its art of web-spinning potential in abstract form to entrap them (insects) in their silk-woven sticky web. Once the prey becomes entangled in the sticky glued web, the spider rushes its prey and injects its venom to immobilize and kill it followed by liquefying its body mass using its gut-wrenched enzyme and sucking the liquid form of proteinaceous body sap of the prey to gain energy (Walter et al., 2017). The amazing feature of the web is that the sticky silk is responsible for gluing down the insects, but no spider has ever been observed entangled in its own web, though there is an intermittent secretion of oily substance from its limb. The chasing and immobilizing of prey insects require proper assessment of the distance and the directional axis of the prey from the spider. As most spiders have poor vision, they cannot judge the location of their prey by visual probing. Rather, they are accustomed to receiving vibrations using their body surface (mainly the sensory reception of skin) and locating the epicenter of vibrations to reach their prey. This trick is closely connected to the fishing lines trick used by fisheries and anglers to catch fish. Though anglers cannot see fish trapped on the hook submerged in water, they can sense the vibration, which helps them catch their target and further releases a signal as an indication for the angler to turn the wheel to get the fish from the water. To maximize the number of catches, a spider weaves several webs at different angles in its prey-hunting zone and spins all those webs in a way for it to easily move from one web to another by maximizing its eight-leg-driven hydraulic movement to get closer to its prey.

A different group of spiders adopts different strategies to catch their prey. It has been observed that garden spiders usually prepare a funnel-web to attract insects, and once they enter, they never come out. The Wolf spiders are fast-moving spiders that usually chase prey insects, catch them, and eat them (Szalay, 2014). Trap-door spiders, meanwhile, usually spin a web in a ground-hole so that a large number of insects can easily become trapped so it can eat them over time (Wegner, 1978). It's quite amazing to witness the multiple strategies that a diverse group of spiders employs to trap insects. One thing worthy of mention is that after infusing its venom in prey to immobilize and kill it, spiders puke the digestive enzymes (e.g., carboxylpeptidase A, aryl aminopeptidase-isozymes, and amidase isozymes) stored in their gut area. They spit it on the prey to let the enzymes digest the soft tissues and liquefy them from inside out. Predator spiders suck the entire liquefied flesh of the prey and leave the outer body shell for other animals to devour (Mommsen, 1978; Walter et al., 2017).

Spiders have an in-built filtration system that puts a check on the solid food to be strained out and throws it from the body to keep it from entering the gut system, which could stop the normal functioning of its digestive system (Bensoussan et al., 2018). Arachnologists have noticed that being predators, most spider species are engaged in hunting insects, whereas many juvenile and small-sized spiders have been observed obtaining their food from plant sap, which can be considered nectar, and scientists have further ascertained that consumption of plant parts helps smaller spiders obtain nutrients and become immunologically strong until their juvenile stage passes (Pappas, 2016). Scientists have noticed some behavioral attributions among spiders based on their food habits, as most insectivorous and araneophagic spiders are hyperactive in the nighttime, although they sometimes hunt during the day, so they are considered quasi-nocturnal (Benson and Suter, 2013).

Spiders often feed on dead animals, mainly dead insects. Nectar-feeding, plant-feeding, and mushroom-eating spiders are agile in broad daylight (Pappas, 2016). Nectar-feeding spiders have been observed to live longer than typical insectivorous and araneophagic spiders, as those animal-eating spiders engage in forcefully puking the digestive enzymes stored somewhere in their gut region on their prey. The continuous production of digestive enzymes forces the spider to spit it on the prey to immobilize or kill it, and this process is highly energy driven and is associated with intermittent cell functioning that supposedly enhances the aging process and concretely shortens its life span in comparison with its nectar-feeding counterparts. It's quite interesting that spiders don't leave their old web and start spinning a new one until they roll up the old one and convert it into a ball-like structure that they also suck on (Warner, 2016). It has also been observed that spiders, along with their old web, eat the dry skeletons of the prey entangled in it, including their molt, to extract nourishment from the biodebris. The residual of the biological wastes, exoskeleton, etc. helps reduce the smell of the organic parts of the prey. The reduction of smell and dead organic wastes reduces the chance of enticing predators to its web, and that might enhance the hunters to face the potential threat to be hunted by their predators. Arachnologists have observed during their studies on the feeding spree of spiders that there is nothing like their nonstop hunting. It's just about to meet its certain benchmark of appetite and then stop feeding. When it feeds itself on a diverse range of insects or dead animals, it keeps on eating whatever is available to it and stores extra energy so it can use it (energy) for difficult periods, as spiders must often survive for several days without food.

From the standpoint of predator—prey dynamics, it's interesting that the predator—prey relationship is not static and straightforward but rather complicated, as scientists observed in Luzon, Philippines. There is an araneophagic spider called the "Web-building spitting spider," *Scytodes* sp., that belongs to Scytodidae and normally hunts its prey named the "Jumping spider," which belongs to Salticidae (Jackson and Pollard, 1996). Arachnologists have determined that *Portia labiatae*, aka the "Aggressive-mimic jumping spider," belongs to Salticidae and has evolved as a result of regional adaptation as a superaraneophagic spider that successfully hunts the dangerous, araneophagic prey "Web-building spitting spider" (Jackson and Pollard, 1996). In fact, localized adaptive radiation is often a cardinal factor in the foraging of predators and act as a determining factor in defining and redefining the predator–prey equilibrium over time.

Another species of *Portia*, the spider belonging to Salticidae, is *Portia fimbriata* in Queensland, Australia, renowned for being a unique example of interspecific araneophagic taxon, which adopts a special strategy to lure *Euryattus* sp., an interspecific arachnid member belonging to Salticidae (Jackson and Wilcox, 1990). The intensive experimental observation of arachnologists was that the female members of *Euryattus* sp. build their nests in suspended, rolled-up leaves and are lured by the vibratory displays of *P. fimbriata*, their potential predator, mimicking the courtship vibratory plays of the male members of *Euryattus* sp. The female members of *Euryattus* sp. usually take the vibratory displays of *P. fimbriata* as the courtship vibratory displays of their male counterparts, i.e., *Euryattus* sp., and this accidental closeness to the predator brings their death. It's quite amazing to see that such behavioral adaptation of *P. fimbriata* in Queensland helped this particular species evolve as an interspecific araneophagic arachnid member. So this is a unique opportunity to see that the evolutionary arms race is a continuous process that defines and redefines evolutionary adaptations and counteradaptations to determine the dynamics of predator–prey equilibrium on this Earth.

According to recent investigations by Swiss scientists, the collective weight (biomass) of 45,000–50,000 globally distributed species of spiders is around 25 million metric tons, and they kill and feed on around 400–800 million metric tons of prey annually (Nyffeler and Birkhofer, 2017). Of the total prey, greater than 90% are insects and collembolans, and scientists have further explored energy flows in a trophic dynamic, and the annual feeding potential of spiders is around 1% of the net primary production of the global terrestrial ecosystem. In their meticulous investigations, the team of Martin Nyffeler and Klaus Birkhofer, University of Basel, Switzerland, noticed that the annual consumption of a diverse group of spiders, part of a diverse ecological niche, varies substantially, as spiders inhabiting forest, woodland, and grassland ecosystems consume >95% of the annual prey they kill. Spiders inhabiting croplands and orchards are reported to consume around 2% of the annual prey they kill, whereas spiders in other habitats like deserts or the Arctic tundra and habitats with anthropogenic interference show a negligible record of annual prey consumption.

If we make qualitative comparisons among spiders and the largest mammals to understand gross annual food consumption by spiders, the results are interesting, as annual consumption by whales and other mammals of the oceanic ecosystem is somewhere between 280 and 500 million metric tons, which is substantially less (Yodzis, 2001) than that of spiders. Going a step further, the Swiss science duo has noted that the recent food consumption record published by the United Nation's Food and Agriculture Organization (Bruinsma, 2003; FAO, 2014) on global food consumption of meat and fish is around 400 million metric tons, in comparison with the annual consumption by spiders, which is in the range of 400–800 million metric tons of biomass. The critical observations on the predator–prey dynamics of spiders and insects led Martin Nyffeler to comment, *"Our calculation status quantifies for the first time on a global scale that spiders are major natural enemies of insects .... In concert with other insectivorous animals such as ants and birds, they help to reduce the population densities of insects significantly .... Spiders thus make an essential contribution to maintaining the ecological balance of nature."* (Nyffeler and Birkhofer, 2017).

References

Agnarsson, I., Coddington, J., Knoflach, B., 2009. Morphology and evolution of cobweb spider male genitalia (araneae, theridiidae). J. Arachnol. 35, 334–395. https://doi.org/10.1636/SH-06-36.1.

Altherr, R., 2019. Leg Uses Hydraulics and Muscle Flex. https://asknature.org/strategy/leg-uses-hydraulics-and-muscle-flex/.

Benson, K., Suter, R., 2013. Reflections on the tapetum lucidum and eyeshine in lycosoid spiders. J. Arachnol. 41, 43–52. https://doi.org/10.1636/B12-35.1.

Bensoussan, N., Zhurov, V., Yamakawa, S., O'Neil, C.H., Suzuki, T., Grbić, M., Grbić, V., 2018. The digestive system of the two-spotted spider mite, tetranychus urticae koch, in the context of the mite-plant interaction. Front. Plant Sci. 9, 1206. https://doi.org/10.3389/fpls.2018.01206.

Britt, R.R., 2008. Creepy Cannibalism: Why Female Spiders Eat Mates. https://www.livescience.com/7555-creepy-cannibalism-female-spiders-eat-mates.html.

Brodie III, E.D., Brodie Jr., E.D., 1999. Predator-Prey Arms Races: asymmetrical selection on predators and prey may be reduced when prey are dangerous. Bioscience 49 (7), 557–568. https://doi.org/10.2307/1313476.

Bruinsma, J., 2003. World Agriculture: Towards 2015/2030, an FAO Perspective. Earthscan, FAO, Rome, Italy.

Cardoso, F.C., Lewis, R.J., 2019. Structure-function and therapeutic potential of spider venom-derived cysteine knot peptides targeting sodium channels. Front. Pharmacol. 10, 366. https://doi.org/10.3389/fphar.2019.00366.

Craig, C.L., Freeman, C.R., 1991. Effects of predator visibility on prey encounter: a case study on aerial web weaving spiders. Behav. Ecol. Sociobiol. 29, 249–254. https://doi.org/10.1007/BF0016398.

Craig, C.L., Weber, R.S., Bernard, G.D., 1996. Evolution of predator–prey systems: spider foraging plasticity in response to the visual ecology of prey. Am. Nat. 147, 205–229. https://doi.org/10.1086/285847.

Durrett, R., Mayberry, J., 2010. Traveling waves of selective sweeps. Ann. Appl. Probab. 21 (2), 699–744. https://projecteuclid.org/euclid.aoap.

Earthkind, 2017. What Do Spiders Eat – the Diet of a Common House Spider. https://www.earthkind.com/blog/what-do-spiders-eat-the-diet-of-a-common-house-spider/.

Edmunds, M., 1974. Defence in Animals: A Survey of Anti-predator Defences. Longman, p. 357.

Edwards, L., 2011. Spider Is the Second Most Vibration-Sensitive Creature. https://phys.org/news/2011-10-spider-vibration-sensitive-creature.html.

Erict_Culinarylore, 2016. Do People Eat Spiders as Food? https://culinarylore.com/food-culture:do-people-eat-spiders-as-food/.

FAO (Food and the Agriculture Orgaization of the United Nations, 2014. Sustainability of Food and Agriculture Systems [User Manual Version 2.2.40). http://www.fao.org/3/a-i4113e.pdf.

Geyer, T., 1967. Polymorphism in the spider, theridioovatum. Nature 213, 314. https://doi.org/10.1038/213314a0.

Jackson, R.R., Pollard, S., 1996. Predatory behaviour of jumping spiders. Annu. Rev. Entomol. 41, 287–308. https://doi.org/10.1146/annurev.en.41.010196.001443.

Jackson, R.R., Wilcox, R.S., 1990. Aggressive mimicry, prey-specific predatory behaviour and predator-recognition in the predator-prey interactions of *Portia fimbriata* and *Euryattus* sp., jumping spiders from Queensland. Behav. Ecol. Sociobiol. 26, 111–119. https://doi.org/10.1007/BF00171580.

Li, D., 2005. Spiders that decorate their webs at higher frequency intercept more prey and grow faster. Proc. R. Soc. B 272, 1753–1757. https://doi.org/10.1098/rspb.2005.3160.

Michálek, O., Kuhn-Nentwig, L., Pekar, S., 2019. High specific efficiency of venom of two prey-specialized spiders. Toxins 11, 687. https://doi.org/10.3390/toxins11120687.

Mommsen, T.P., 1978. Digestive enzymes of a spider (Tegenaria at rica Koch)—I. General remarks, digestion of proteins. Comp. Biochem. Physiol. Physiol. 60 (4), 365–370.

Nyffeler, M., Birkhofer, K., 2017. An estimated 400–800 million tons of prey are annually killed by the global spider community. Sci. Nat. 104. https://doi.org/10.1007/s00114-017-1440-1.

Pappas, S., 2016. 305-Million-Year-Old 'Almost Spider' Unlocks Arachnid History. https://www.scientificamerican.com/article/305-million-year-old-almost-spider-unlocks-arachnid-history/.

Roach, J., 2001. Fear of Snakes, Spiders Rooted in Evolution, Study Finds. https://www.nationalgeographic.com/science/2001/10/fear-evolution-spiders-science/.

Sneed, A., 2014. Fact or Fiction?: People Swallow 8 Spiders a Year while They Sleep. https://www.scientificamerican.com/article/fact-or-fiction-people-swallow-8-spiders-a-year-while-they-sleep1/.

Stevens, A., 2010. Predation, herbivory, and parasitism. Nat. Ed. Knowled. 3 (10), 36.

Szalay, J., 2014. Wolf Spiders: Bites, Babies & Other Facts. https://www.livescience.com/41467-wolf-spider.html.

Virant-Doberlet, M., Kuhelj, A., Polajnar, J., Šturm, R., 2019. Predator-prey interactions and eavesdropping in vibrational communication networks. Front. Ecol. Evol. 7. https://doi.org/10.3389/fevo.2019.00203.

von May, R., Biggi, E., Cárdenas, H., Diaz, M.I., Alarcón, C., Herrera, V., Santa-Cruz, R., Tomasinelli, F., Westeen, E.P., Sánchez-Paredes, C.M., Larson, J.G., Title, P.O., Grundler, M.R., Grundler, M.C., Rabosky, A.R.D., Rabosky, D.L., 2019. Ecological interactions between arthropods and small vertebrates in a lowland Amazon rainforest. Amphib. Reptile Conserv. 13 (1), 65–77 (e169).

Walter, A., Bechsgaard, J., Scavenius, C., Dyrlund, T.S., Sanggaard, K.W., Enghild, J.J., Bilde, T., 2017. Characterisation of protein families in spider digestive fluids and their role in extra-oral digestion. BMC Genom. 18 (1), 600. https://doi.org/10.1186/s12864-017-3987-9.

Warner, D., 2016. Amazing Quality Cobwebs. https://www.irishexaminer.com/opinion/columnists/arid-20414576.html.
Wegner, G.S., 1978. Spider Identification Guide. BASF-The Chemical Company, p. 60. Spider-Guide-Wegner-BASF-Revised-12-2-14 (2).pdf. www.researchgate.net.
Ximenes, N.G., Gawryszewski, F.M., 2019. Prey and predators perceive orb-web spider conspicuousness differently: evaluating alternative hypotheses for color polymorphism evolution. Curr. Zool. 65 (5), 559–570. https://doi.org/10.1093/cz/zoy069.
Yong, E., 2016. The Bug that Stalks Spiders on Their Own Webs. https://www.theatlantic.com/science/archive/2016/10/the-bug-that-stalks-spiders-on-their-own-webs/505280/.
Yong, E., 2017. The Spider Webs that Gets Stronger when it Touches Insects. https://www.theatlantic.com/science/archive/2017/05/crebillate-spider-web/528585/.
Yodzis, P., 2001. Must top predators be culled for the sake of fisheries? Trends Ecol. Evol. 16, 78–84.
Zhang, S., Koh, T.H., Seah, W.K., Lai, Y.H., Elgar, M.A., Li, D., 2011. A novel property of spider silk: chemical defence against ants. Proc. R. Soc. B 279 (1734), 0962–8452. https://doi.org/10.1098/rspb.2011.2193.

CHAPTER SIX

Evolutionary and Ecological impacts of stenophagy in Araneae: Its role in promoting or limiting spider diversity on planet Earth

Abstract

Ecologists and evolutionary biologists have elucidated stenophagy (narrow diet breadth and prey specificity) as an ultimate level of ecological (trophic) and evolutionary (phylogenetic) specialization witnessed among animals such as spiders, which are prehistoric terrestrial carnivores. Contemporary investigations have ascertained that stenophagy emerged independently on several occasions. Arachnologists have witnessed that within families, trophic traits such as euryphagy and oligophagy are in basal condition, whereas stenophagy is the derived state. Stenophagy in spiders broadly falls within six categories: araneophagy, crustaceophagy, dipterophagy, lepidopterophagy, myrmecophagy, and termitophagy. Investigations into the evolution of stenophagy in spiders have led scientists to wonder whether it promotes diversification or extinction of spiders and whether it is a phylogenetically conserved and evolutionary derived state. The evolution of stenophagy in Araneae is regulated by a combination of ecological and phylogenetic factors.

Keywords: Araneae; Predator; Prey; Prey specificity; Stenophagy.

On the basis of 1115 records of Evarcha culicivora feeding in the field, we can characterize this East African jumping spider (Salticidae) as being distinctively stenophagic.... Mosquitoes were the dominant prey (80.2% of the records), with the majority (82.9%) of the mosquitoes being females, and thereafter midges were the most common prey (9.2% of the records). Preference profiles that were determined from experiments showed strong convergence with natural diet in some, but not all, instances.

Robert R. Jackson and Others.

Most animal biologists, animal taxonomists, ecologists, and arachnologists would like to recognize spiders and Araneae (the seventh diverse order within the animal kingdom) exclusively as terrestrial carnivores (an animal

Spider Evolution
ISBN: 978-0-323-90041-6
https://doi.org/10.1016/B978-0-323-90041-6.00010-3

that eats only animal material). This desired classification might be because they did not observe the herbivorous (animals that eat plant materials) feeding habits of these creatures (Coddington and Levi, 1991). But ecologists have observed a diverse number of tree-dwelling spiders (mainly living in *Acacia* trees) feeding on nectars from their host trees. Furthermore, the feeding habit of *Bagheera kiplingi*, a spider species from Central America, which has a herbivore (Meehan et al., 2009), has again led scientists to recognize these fearsome creatures as terrestrial omnivores (animals that eat both plant and animal materials) rather than as a predominantly terrestrial carnivore. Based on a series of experiments and paleontological and phylogenomics reviews, evolutionary biologists have ascertained that Araneae is the maiden predator to evolve within the Earth's terrestrial ecosystem (Pisani et al., 2004). In this context, we should recognize these animals as predators, as they usually eat living things. But to present a concise format, we recognize as predators those animals that typically chase and hunt prey. It has been observed that predation could restrict the diversity of prey species by controlling their populations. Ecologists have recognized that the extent of predation steers natural selection toward evolutionary stability by involving several factors during predation such as mimicry, change in exterior color, nesting strategies, and alteration of sexual selection.

In general, however, because they evolved as the earliest terrestrial predator, most web-spinning smaller-sized spiders trap a diverse group of tiny insects such as flies, mosquitoes, and moths on their sticky webs before feeding on them. On the other hand, large, exotic spiders have been observed devouring diverse prey, including small birds, fish, reptiles, toads, and small mammals (such as rodents). Immediately after capturing their prey, carnivorous spiders inject venom into them and accomplish the course of external digestion in two distinct ways:

a. Several carnivorous spiders pump digestive enzymes into their prey, which dissolves the solid and semisolid internal tissues into liquid form and leaves the prey's exoskeleton as a husk. It helps predatory spiders suck the liquid tissues to store them in their stomachs.

b. Another group of predatory spiders grinds and mechanically scavenges the solid tissues of prey with their fangs (chelicerae), the mouthparts or jaws, and the pedipalp (the lateral extension of chelicerae), followed by injecting a stream of enzymes to liquefy the remaining smaller solid tissues, by mechanical scavenging, before feeding on them.

Let's try to get acquainted with the philosophical introspection of the origin and evolution of feeding habits in animals that vary by feeding strategy and have evolved several times independently on Earth. From an ecological and evolutionary perspective, scientists have considered that the emergence of carnivorous as a transitional state from insectivorous, for the medium and large-sized tetrapods, must go through the necessary exposition of adaptation. In support of their observations, evolutionary biologists have further claimed that in the evolutionary history of the emergence of terrestrial tetrapods, large-sized amphibians have emerged out as piscivores around 400 mya. Eventually, the amphibians continued their traditional feeding habit along with negligible diversification of their food habit. They did so by starting the consumption of a large number of fish from bodies of water and by exploring new feeding habits with the consumption of reptiles and insects and smaller tetrapods and consumption of plant materials. Eventually, it leads to the evolution of carnivorous and herbivorous food habits in tetrapods, which emerged initially as piscivorous (Sahney et al., 2010). Thus, it is a proven fact that the evolution and diversification of feeding strategies have made it possible for the extant group of animals to be diversified immediately at the microevolutionary level. In addition, it also streamlined the ultimate journey of biological evolution to its macroevolutionary level, where the tetrapods, with their diverse food habits, occupied almost all terrestrial ecosystems on Earth as a predominant animal group within the animal kingdom during this era. The feeding habits of the animals fall into two categories:

1. Monophagy: when animals feed on only one type of food—for instance, insects feeding on a specific plant species of a specific family.
2. Polyphagy: when animals feed on more than one type of food—for example, insects feeding on a diverse number of plants within the same families.

   In addition to these two distinct categories, scientists have recognized a third category, referred to as oligophagy:

3. Oligophagy: when animals feed on a limited number of foods restricted to a specific taxonomic group—for example, pests that grow on leguminous host plants and feed on them.

Whether it is a monophagous or polyphagous kind of food habit of animals, it seems that a simple behavioral trait likely appeared as a random choice that favored one over another. But it is a fact that the evolution of a distinct type of feeding habit, evolved in independent lines, would steer the natural selection to a certainty. This is because it has appeared to be the decisive factor in recognizing particular animals as predominant and diversified in the evolutionary journey of the animal kingdom. In support of this observation, scientists have argued, "*An insect's choice of a food plant is assumed to be based on a simple decision rule that maximizes the expected production of offspring. Monophagy or polyphagy may be favored, depending on the proportion of an extended diet that would be unsuitable if chosen, versus the difficulty in finding the most suitable food*" (Levins and MacArthur, 1969). However, as far as the air-breathing arthropods (with eight legs, two chelicerae with fangs to inject venom in hunting prey, and unique web-spinning potential) are concerned, spiders (of the order Araneae) are considered the seventh-largest diverse order within the animal kingdom in terms of species diversity (Selden et al, 2008; Sebastin and Peter, 2009). Though several systems of classification have been in practice since 1900, the phylogenetic system of classification in 2021 estimated 49,800 spider species (in 129 families) within the terrestrial ecosystems of Earth [World Spider Catalog, Natural History Museum, Bern, retrieved July 17, 2019] (Kropf et al, 2019). Paleontologists have determined that "spiderlike arachnids" evolved around 386 million years ago (mya), during the Devonian Period. They have evolved with silk-producing spigots, but scientists did not find any spinnerets among them (Selden et al., 2008).

Down the evolutionary trail of history of origin and evolution of spiders, the meticulous examinations of the fossilized spiders trapped in the Carboniferous rocks (between 318 and 299 mya) have shown that the transitional, primitive spiders belong to the primitive but extant order Mesothelae (Garwood et al., 2016). Further, scientists have observed that the emergence of the modern group of Mygalomorphae and Araneomorphae spiders evolved around 200 mya around the Triassic Period (Starrett et al., 2012). Further, it has been claimed that the emergence of predominant terrestrial carnivores in most inhabitable terrestrial ecosystems on Earth is an ecological favor due to the inherent, diverse nature of trophic strategies that most spiders are endowed with since their maiden appearance on Earth (Coddington and Levi, 1991). Arachnologists have observed a diverse array of trophic strategies ranging from euryphagy to oligophagy to stenophagy, but they have not recognized any typical monophagous spider in the order Araneae. Thus, in the evolution of the origin and diversification of many spider species as well as their existence in diverse terrestrial ecosystems, acquiring and employing inherent, varied trophic strategies is an essential factor and plays a crucial role in maintaining evolutionary dynamics and regulating the predator—prey dynamics of spiders and their prey. The major trophic strategies of spiders are as follows:

a. Euryphagous: Most arachnologists have recognized spiders involved in chasing and hunting a sizeable crowd of invertebrate prey for their platter as "euryphagous" (Nentwig, 1987).

b. Oligophagous: Arachnologists have not recognized a monophagous spider, although they have recognized a few species involved in capturing and devouring specific groups of prey. However, they do not depend on their limited prey for food; rather, they supplement their diets by hunting alternative prey. For example, the salticid *Portia* spp., which typically hunt and feed on other spiders, hunt and feed on insects when spiders are absent (Li et al., 1997).

c. Stenophagous: Stenophagy or narrow diet breadth in spiders is a unique trophic specialization and is a derived state from an evolutionary perspective, which has emerged among the primitive, terrestrial carnivores on Earth, like spiders. Monophagous food habits are typically not observed among spiders, and the act of oligophagy does not appear so often among them. Yet critical reviews on the trophic strategies of spiders have found that some spider species have evolved with a narrow food habit due to the availability of a single type of prey (which is technically monophagy) or due to the restricted availability of prey (which is oligophagy). This emergence of a derived state of food habit among spiders with a narrow breadth diet seems to be an instance of a unique trophic specialization and is considered the ultimate driving force behind the evolutionary journey of spiders. For example, arachnologists have referred to the zodariids, *Zodarion* spp., which have employed myrmecophagy (a specialized ant-eating food habit) to capture and consume only ants (Pekár, 2004). From the perspective of ecological strategies, as far as the classification of diet in animals is concerned, scientists have broadly recognized the existence of two extreme extents of "trophic breadth": stenophagy and euryphagy. It was surprising for ecologists to determine the inherent forces leading the evolution of such ecological strategies, and it might further lead to the recognition of two distinct extremes, "specialist diet" and "generalist diet" (Levins and MacArthur, 1969).

"Stenophagy" is a derived state, a typical state of ecological specialization, as stenophagy is involved in trade-offs, having a unique depiction in the form of the interaction between predators and their nonpreferred prey (Joshi and Thompson, 1995; Fry, 1996). Evolutionary biologists have contemplated that such trade-offs affect the evolutionary direction and alter the speciation rate of the concerned organism engaged in this process.

Hence, it has been further construed that a transitional mode from euryphagy to stenophagy is more likely than the reverse process. From this, scientists understand that stenophagy is a specialized trophic strategy to enhance the multitude and extent of extinction of concerning organisms. Ecologists have observed that stenophagy in those animal taxa occupies a narrow trophic niche and these animals are predominantly herbivores and parasites, but it is also found among carnivores (Jermy et al., 1990; Poulin, 1992; Hodek and Honek, 1996). Though the emergence of dichotomously evolved diet, such as "specialized-generalized," had steered the microevolutionary journey of organisms such as arthropods or spiders toward a certain evolutionary trail, it either pushed them to evolve into the diversified mode or pushed them to the brink of extinction against the backdrop of the evolutionary timescale. Ecologists have generally observed that a transition from a generalist to specialist diet, one that is very common in insects and amphibians (frogs); though the transition from specialist to generalist strategies is also witnessed among parasitoids (Gilbert et al., 1994; Nosil, 2002; Stireman, 2005). The attributes of stenophagous predators are mostly unique physiological, biobehavioral, and morphological traits that allow them to efficiently use those characteristics for hunting their specific choice of prey rather than hunting a wide range of prey. Ecologists have contemplated that such a transition from a generalist trophic strategy to a specialist trophic strategy must have an evolutionary basis defining the emergence of this trophic specialization (stenophagy) in terrestrial arthropods, specifically the evolution of terrestrial carnivores within the diverse terrestrial ecosystems on Earth.

Let's briefly review some hypotheses to help us better understand the evolution of stenophagy, especially among herbivores, for which evolutionary biologists and ecologists have short-listed several variables (Singer, 2008):

1. Coevolutionary relations between prey and predators and trade-offs
2. Optimum foraging that regulates predator—prey dynamics
3. Neural constraints
4. Enhanced physiological potential of predators
5. Use of enemy-free space during the foraging of predators
6. Escaping interspecific competition to become involved in optimizing foraging and other tasks

Whenever investigators have studied the evolutionary role of the emergence of stenophagy in the life cycle of several invertebrates and vertebrates, the narrow diet breadth has played a cardinal role in the macroevolutionary process (Darst et al., 2005). Additionally, whenever any organism would have tried to adapt to a new trophic zone (e.g., a transitional move of frogs from water bodies to the terrestrial ecosystem), it would result in the evolutionary edge in the diversification of predators (Futuyma, 1986). But there is still controversy within the arena of science to claim unequivocally whether the species diversification extends to stenophagous lineages or euryphagous lineages, as many investigators firmly affirmed that the stenophagous lineages are less diversified than the euryphagous ones (Toft, 1995). On the other hand, some ecologists have observed that stenophagous specialization supposedly enhances the rate of diversification of species by cutting down the gene flow in maintaining the equilibrium state of a population (Futuyma and Moreno, 1988).

To explore the phylogenetic and ecological correlations with stenophagy in a diversified organism, specifically, any diverse group having a large clad, Dr. Stano Pekár, from Masaryk University, Czech Republic, along with two other research colleagues, Jonathon A. Coddington, from Smithsonian National Museum of Natural History, USA and Todd Blackledge, from the University of Akron, USA, has engaged in investigating the evolutionary role of stenophagy on spiders and determining whether

a. Stenophagy has promoted the diversification of species
b. Stenophagy is a phylogenetically conserved and evolutionary derived state
c. Stenophagy is determined by geographical distribution and foraging guild

In the format of secondary research, Pekár and others (2011) have studied the published data of 600 spider species. They have initially witnessed six types of stenophagy in those spiders:

1. Myrmecophagy
2. Araneophagy
3. Lepidopterology
4. Termitophagy
5. Dipterophagy and
6. Crustaceophagy

Though the mechanism of steering the origin and evolution of stenophagy in spiders and the resulting implications of its speciation are missing in contemporary ecological investigations (Pekár et al., 2011) have presented unique findings. This includes the observation that the diets in diverse spider species have evolved and diversified over the last 100 years with phylogenetic progress at distinct taxonomic levels. It has been further ascertained that it was expected that the evolutionary diversification of spiders supposedly follows the model of evolutionary diversification of arthropods. This is because the major trophic shifts in arthropods play a very important role. But it did not happen in the case of spiders. Rather, the evolutionary diversification of spiders is related to their use of web-spinning potential, which ultimately helps them catch diverse prey insects (Bond and Opell, 1998). In the context of evolution and diversification of the web-spinning potential of spiders, arachnologists have observed that spiders used a different kinds of silk threads (oozed from different silk glands) to form structurally different types of webs, suitable for a distinct type of microhabitat (Blackledge et al., 2009). Some structurally and functionally, webs are unique in capturing their specific prey (Bond and Opell, 1998; Blackledge et al., 2009). Further, most orb-spinning spiders are using their webs to catch their prey. On the other hand, the spider species such as *Mastophora* sp., *Ordgarius* sp., and *Cladomelea* sp. (of the Araneidae) hunts their prey with extremely reduced webs and are capable of hunting specific types of prey such as moth species (Yeargan, 1994). This prompted evolutionary biologists and arachnologists to hypothesize that the origin and evolution of prey specialization have supposedly taken place in the cursorial spider species, as they do not use silk webs to catch and consume their prey, in comparison with the web-spinning spiders that are less selective in their diet (Foelix, 1996).

Stano Pekár and his colleagues (2011) have primarily investigated the phylogenetic pattern of stenophagy among spiders and tried to determine what type of stenophagy is predominant. In the next level of their investigation, they have tried to test a couple of hypotheses—(1) diversification hypothesis and (2) phylogenetic constraint hypothesis—to understand how diet breadth (specifically the narrow diet breadth) influences the evolutionary diversification of spiders.

a. Diversification hypothesis: According to the "diversification hypothesis," the diet breadth of the stenophagous family differs from that of the euryphagous family and genera of any given organism. From the ecological investigations, it has been witnessed that stenophagy has substantially decreased the extent of species diversity, and that might enhance the extent of species extinction in the long run.

b. Phylogenetic constraint hypothesis: According to the "phylogenetic constraint hypothesis," either stenophagy or euryphagy is considered a derived condition. However, this is restricted in a few clades of spiders. Additionally, the researchers have presented the "Ecological Constraint Hypothesis" in two different forms, and the leading group of research investigators has claimed that "diet breadth differs among geographical regions."; whereas the other group has observed that "diet breadth differ between foraging guilds" (Pekár et al., 2011).

To unfurl the ecological and evolutionary importance of the emergence of stenophagy in spiders, Pekár et al. (2011) have picked spider species, and the diet breadth of each one of them has been computed, and the heuristic index (H) has been chosen, using the Shannon–Wiener Index (Weaver and Shannon, 1949). It is determined that the range of the diversity index is between 0 and 3. In their experimental model, they have defined the heuristic threshold limits for stenophagy (H = 0–0.3), oligophagy (H = 0.31–1.1), and euryphagy (H = 1.11-3). Using these defined formats of heuristic threshold limit, they have identified that there are 156 stenophagous, 144 oligophagous, and 262 *Euryphagus* species of spiders in their experimental dataset.

To test the "diversification hypothesis," Pekár et al. (2011) have chosen euryphagous and stenophagous genera and families and compared their diet breadth. They have specifically examined the families and genera of those spider species, the index of prey diversity of which were (mean of H for species in a genus/family < 0.3) in the threshold limit for stenophagy, and the number of prey was restricted to 10. There were 6 families and 45 genera in this experimental dataset. In the next level, for determining each stenophagous family/genus, they have screened out the closely related sister family/genus for those spiders, and as a result, the evidence of euryphagy (mean of H for species in a genus/family > 1.1) was readily considerable. Finally, Pekár et al. (2011) were engaged in the comparison of the number of the species for a stenophagous family/genus with that of the nearest *Euryphagus* family/genus conducted by Wilcoxon tests (Mitter et al., 1988). stenophagous/euryphagous diversification, sister family comparisons were made in the species that belonged to the following family pairs: Archaeidae/Eresidae, Ammoxenidae/Gnaphosidae, Caponiidae/Segestriidae, Mimetidae/Eresidae, Oonopidae/Dysderidae, Palpimanidae/Eresidae, Zodariidae/Amaurobiidae.

To test the "phylogenetic constraint hypothesis," Pekár et al. (2011) have constructed phylogenetic trees by compiling the phylogenetic and taxonomic data, publishing and assuming that branch distances were the same as those data that were not yet reported. There is a gray zone in understanding from a phylogenetic perspective that whether the emergence of stenophagy in spiders should be considered basal or as a derived state because both promulgations have been substantiated by several hypotheses. To sort this out, the researchers have critically analyzed the correlation between prey diversity and the branch distance of the family tree. But these two factors were not correlated, leading scientists to contemplate that the origin of stenophagy in spiders should be regarded as the derived state rather than the basal state. Pekár and his colleagues have started with phylogenetic data exploration for Araneae, and they did not find adequate phylogenetic information for the newly added families such as Cybaeidae, Hahniidae, Homalonychidae, and Nephilidae (Coodington, 2005). The phylogenetic positions have been sorted out from the analysis of recent works of Jocque and Sppenaar-Schoeman (2007). In their experimental model, Pekár et al. (2011) has outlined the formation of generic level phylogenies for each family:

1. Having four genera (having data on its prey)
2. Having more than one stenophagous species so that they could yield a phylogenetic analysis.

The research investigators have short-listed around eight families of Araneae having available phylogenetic info that could fit in the criteria mentioned above):

i. Araneidae (Scharff and Codington, 1997; Agnarsson and Blackledge, 2009).
ii. Corinnidae (Bosselaers and Jocqué, 2002)
iii. Dysderidae (Arnedo et al., 2007b)
iv. Salticidae (Maddison and Hedin, 2003; Maddison et al., 2008)
v. Theridiidae (Arnedo et al., 2004, 2007a)
vi. Thomisidae (Benjamin et al., 2008)
vii. Tetragnathidae (Alvarez-padilla et al., 2009) and
viii. Zodariidae (Jocqué, 1991).

To analyze their experimental model, the generalized least squares (GLS) model was used to examine the correlation between the branch distance and the "continuous response variables" (H) at the family level. To carry out these phylogenetic and evolutionary analyses, researchers have used generalized estimating equations (GEEs) with binomial errors, which were used earlier in the "Ape package" (Paradis, 2006). A relationship was set between the branch distance and the proportion of prey in the diet of a potential predator (which is the response variable H). The researchers have observed that the low diversity index empirically indicated the state of stenophagy, and it is considered the outcome of a relationship dynamics between a potential predator having an extremely high proportion of certain prey (Pekár et al., 2011). Based on the preliminary observations, it was interpolated that an empirical result was desired, when "*the relationship between the high proportion of certain prey and branch distance would decrease linearly on a logit scale if stenophagy was a basal condition, increase linearly, if stenophagy was a derived condition, or be quadratic*" (Pekár et al., 2011). The researchers have tried to determine the branch distances based on the number of nodes, supposedly separating each species from the root of the tree (Stireman, 2005). Both phylogenies have been analyzed that included all taxa and reduced phylogenies and all available prey data. It was observed that both yielded almost identical data, and consequently, the data on the reduced phylogenies have been taken into account (Pekár et al., 2011). Furthermore, the researchers have engaged in the estimation of ancestral states for continuous characters (prey diversity and prey proportion) for each node of the eight families mentioned above. It has been estimated at the genus and species level, "using maximum likelihood, assuming Brownian motion within the ace function from ape package" (Schlutter et al., 1997), and it was observed that ancestral state estimation works do not allow for weighting, yet yields estimate for each node.

The researchers have reviewed the two hypotheses (diversification hypothesis and phylogenetic constraint hypothesis) to investigate the position of stenophagy along the phylogeny, and the estimation was like that of "logit regression using GEE" and "ancestral state estimation using ace." It was observed that testing of both hypotheses yields similar but not identical results. The distribution of euryphagous and stenophagous species of Araneae was made according to the zoogeographical template of Platnick (2010), who has classified the entire geographical regions into four zoogeographical regions: tropical, subtropical, temperate, and cosmopolitan. The Ecological Constraint Hypothesis yielded more precise results than the diversification hypothesis to elucidate the phylogenetic and evolutionary aspects of stenophagy in spiders. In the Ecological Constraints Hypothesis, the researchers have used GLS to compare the prey diversities in different zoogeographical regions and foraging guilds.

Pekár and his research colleagues (2011) have claimed that the species diversity of euryphagous and Stenophagous representatives in spiders in genera and family level are identical, though the emergence of stenophagy (a derived state) has taken place independently and repeatedly. Furthermore, at the family level of these spiders, it has been observed that the basal condition is represented by oligophagous and euryphagous members.

Because Araneophagy is restricted to basal and intermediate lineages, scientists recognized it as an ancestral state. It is quite interesting that

Pekár et al. (2011) witnessed that the diet breadth of spiders was less diverse in spiders inhabiting tropical and subtropical regions relative to their temperate counterparts. Further, the low index of diet breadth has been observed more among the cursorial spiders than in their web-spinning counterparts.

Evolutionary biologists and arachnologists have ascertained that spiders are the prehistoric maiden predators that invaded and occupied terrestrial ecosystems (Pisani et al., 2004). Based on the paleontological and phylogenetic evidence, it is contemplated that the origin and emergence of araneophagy (eating of spiders in the intraspecific and interspecific guild) has supposedly taken place during Devonian, at the time of diversification of insects, and in the race of survival, the diversified group of spiders has engaged in cannibalism to hunt for spiders of same or different species nearby (Vollarth and Selden, 2007). The act of cannibalism in spiders is considered an ancestral state from an evolutionary perspective, and ecologists observed that the spider communities inhabiting at the base of trees have engaged in this act of araneophagy. It was further observed that the extant, cursorial (not the web-spinning species) members of Araneae are araneophagic (Rypstra and Samu, 2005). The trophic specialists further classified this act of araneophagy in the following categories:

1. Sexual cannibalism: When one of the partners cannibalizes its counterpart either before or after copulation (Elgar, 1992).
2. Cannibalism: When an individual consumes another individual of the same species as food, it is considered an act of cannibalism (Polis, 1981).
3. Intraguild predation or IGP: IGP is a sneaky act of hunting and eating of a potential competitor of a different species. This is an interaction between predation and competition, where both of them rely on common prey and proceed to prey on each other to get the benefit (Wise and Chen, 1999).
4. Oophagy: It is an act of embryos feeding on eggs produced in the ovary inside the mother's uterus (Valerio, 1974).

Apart from these four major categories, there is another category in araneophagy, called "kleptoparasitism," which requires a brief insight to understand the evolution of araneophagy. There is a perception that the web-spinning members of Araneae are bestowed with a unique evolutionary trait, where they spun the sticky web to capture and consume their prey, but at the same time, they could move freely on that web. It seems that the spiderweb has assisted its creators in the entrapment and in the consumption of a diverse group of prey effortlessly, but this is not true in real life.

The entomologists and the animal taxonomists have identified a group of spider-hunting bugs (eight-legged predators), commonly known as "thread-legged assassin bugs," of the subfamily *Emesinae*, and family *Reduviidae*, under the suborder Heteroptera (Boll, 2017; Wignall and Taylor, 2010). It is theorized that these "thread-legged assassin bugs" evolved with a special taste for feeding on spiders. Araneophagic assassin bugs employ two unique predatory strategies, "stalking" and "luring" (Wignall and Taylor, 2010). When stalking, assassin bugs approach slowly to get prey spiders within their range. Once they obtain their prey on the range, they rip off the silk threads of the web spun by that spider and hunt. In luring their prey, assassin bugs stretch their legs on the legs of their prey to produce vibrations to communicate with the resident spiders. These resident spiders misjudge the source of vibrations as a particular location on the web and presume that their potential prey is trapped therein, and resident spiders thus rush to the source of the vibration. When spiders are within their range, assassin bugs jump on them to capture and consume them. Occasionally, assassin bugs do not attack the resident spider directly; instead, they wait proximate to the web. Whenever an insect is trapped in the spider's web, assassin bugs immediately rush at that trapped prey, rip off the silk threads of the web to steal and consume it. This kind of sneaky behavior of the assassin bugs toward the web-spinning resident spider is referred to as "parasitism by stealing" or "kleptoparasitism" (Boll, 2017). Some "thread-legged assassin bugs" involved in araneophagy are *Bagauda cavernicola, Eidmannella pallida, Emesa mourei, Phasmatocoris labyrinthicus, Stenolemus bituberus* (Wygodzinsky, 1966; Wignall and Taylor, 2010; Resende et al., 2016). Though araneophagy is considered an ancestral state, so it is supposed to be more frequently witnessed among the basal taxa *Salticidae, Zodariidae,* and *Araneidae*. But a group of arachnologists has categorically denounced the ancestral state of araneophagy as they claimed that araneophagy is homogeneously distributed within the taxa of Theridiidae, Tetragnathidae, Corinnidae, Thomisidae, etc. throughout Araneae.

Evolutionary biologists have discovered that in the age of Permian, the emergence of Crustacean-isopod prey and the Crustacea-eating predator (of the family Dysderidae) have evolved with Crustaceophagy for the first time (Vollarth and Selden, 2007). The several stenophagous crustacea-consuming spiders include the prominent *Amaurobioides* (of Anyphaenidae) and *Tedia* (of Dysderidae).

However, as far as the evolution of stenophagy is concerned, the entomologists have ascertained that its evolution in flying insects emerged in the

Jurassic and Triassic Eras. Further, it is observed that "two-dimensional web-building species" (of Orbiculariae) and the act of dipterophagy (an act of eating Diptera) supposedly evolved in the Jurassic Era (Ayoub et al., 2007). Stenophagous Diptera-consuming spiders include *Cryptachaea*, *Phoroncidia* (of Theridiidae), *Mastophora* (of Araneidae), and *Tetragnatha* (of Tetragnathidae). Likewise, it has been determined that Araneid predators and Lepidopteran prey evolved during the Jurassic Era, and lepidopterophagy has been derived in Araneidae (Grimaldi and Engel, 2005; Ayoub et al., 2007). A few stenophagous Lepidoptera-consuming spiders are *Celaenia*, *Kaira*, *Mastophora*, and *Pasilobus* (of Araneidae), although evolutionary biologists have observed that maiden emergence of termites prey and their potential predators (of Theridiid) took place during the Cretaceous Period, so termitophagy should categorically be recognized as the derived state for termitophagous specialists. But it is not ecologically determined so far due to the absence of enough evidence on the predator–prey dynamics of termitophagous species and their prey (Grimaldi and Engel, 2005; Vollarth and Selden, 2007). Stenophagous termite-consumers include *Ammoxenus* (of Ammoxenidae), *Chrosiothes* (of Theridiidae), and *Stenaelurillus*, or *Microheros* (of Salticidae). From the perspective of myrmecophagy's evolutionary emergence, ants emerged as prey during the Cretaceous. On the other hand, there are the ant-eating Theridiidae and a small number of "non-web-building families" with retrolateral tibial apophysis on the male palp (RTA). The RTA clade of Orbiculariae—*Zodariidae*, *Salticidae*, *Corinnidae*, and *Thomisidae*—are younger than ants (as far as evolutionary aspects) and emerged with myrmecophagous characteristics, which are a derived state. Stenophagous ant-consuming spiders include *Amyciaea*, *Tmarus* (of Thomisidae), *Anasaitis*, *Tutelina* (of Salticidae), *Asagena*, *Steatoda*, *Yaginumena* (of Theridiidae), *Callilepis* (of Gnaphosidae), *Falconina* (of Corinnidae), *Galianoella* (of Gallieniellidae), and *Habronestes*, *Zodarion* and *Zodariellum* (of Zodariidae). Consequently, scientists have tried to interpret the occurrence of stenophagy in a diverse array of spiders. However, this should properly be interpreted because a precise interpretation of convolutions of any specific prey and its predator would be unfurled in the backdrop of spatiotemporal scale.

During their intensive effort to classify the predatory strategies and foraging potential, arachnologists have found a clear dichotomy in the behavioral traits of Araneae: cursorial predator (e.g., *Heteropoda venatoria*) and web-spinning predators (e.g., *Argiope bruennichi*) (Harmer et al., 2011; Zhang et al., 2015). Further, the behavioral trait has been attributed to species specificity, as both traits are not found within any particular species, nor have species with a specific trait been found with another trait. It is quite an interesting experimental observation that of the 663 predatory species of Araneae, the web-spinning guild has captured around 255 prey species in its web, whereas the cursorial guild has captured around 308 prey species (Pekár et al., 2011). The researchers have observed that dipterans are the predominant prey, trapped in the web-spinning guild, whereas small spiders and ants are the predominant prey for cursorial species. Prey such as crustaceans, lepidopterans, and termites are captured by the predators of both types of guilds, and they consume the prey at the same frequencies. With two distinct guild types, two distinct types of radiations have been found to be conspicuous in Araneae. When species diversification is observed in Orbiculariae for the web-spinning guild, the RTA clade is observed for the cursorial guild (Blackledge et al., 2009). It was contemplated that in both guilds, stenophagy evolved (in both prey) from ancestral predatory strategies. The earnest effort of arachnologists in mapping stenophagy in Araneae in tree habitats clearly shows that in *Mygalomorphae*, stenophagy is absent and rarely found within the members of Haplogynae and is predominant among the RTA clade of Orbiculariae (Pompozzi and Copperi, 2018), although Pekár and his research colleagues (2011) traced out around 600 stenophagous species of Araneae. Some published experimental data reveal that 1.4% of the 41,000 spider species, or 574, are stenophagous (Platnick, 2010), a figure that approximates the 600-species observation of Pekár et al. (2011). Finally, arachnologists have concluded that understanding the evolution of stenophagy in spiders requires more ecological investigation of the adaptations of stenophagous species that limit capture and consumption of specific prey (diet specialization) by stenophagous spiders. Furthermore, it is argued that the evolution of stenophagy in spiders has been influenced by ecological and evolutionary variables (specialization of prey, cursorial hunting strategies, etc.), and stenophagy itself should not be regarded as a factor that limits or promotes diversification of any species, including spiders. However, a stereotypic "species-poor" state of the stenophagous spider indirectly supports "a limiting influence of stenophagy on diversification of spiders" (Pekár et al., 2011).

References

Agnarsson, I., Blackledge, T.A., 2009. Can a spider web be too sticky? Tensile mechanics constrains the evolution of capture spiral stickiness in orb-weaving spiders. J. Zool. 278, 134–140.

Alvarez-Padilla, F., Dimitrov, D., Giribet, G., Hormiga, G., 2009. Phy- logenetic relationships of the spider family *Tetragnathidae* (Araneae, *Araneoidea*) based on morphological and DNA sequence data. Cladistics 25, 109–146.

Arnedo, M.A., Coddington, J., Agnarsson, I., Gillespie, R.G., 2004. From a comb to a tree: phylogenetic relationships of the comb-footed spiders (Araneae, *Theridiidae*) inferred from nuclear and mitochondrial genes. Mol. Phylogenet. Evol. 31, 225–245.

Arnedo, M.A., Agnarsson, I., Gillespie, R.G., 2007a. Molecular insights into the phylogenetic structure of the spider genus *Theridion* (Araneae, *Theridiidae*) and the origin of the Hawaiian Theridion-like fauna. Zool. Scripta 36, 337–352.

Arnedo, M.A., Hernandez, M., Batista, B., Oromi, P., 2007b. Mediterranean Biogeography and Island Evolution as Exemplified by the Spider Family Dysderidae. Poster at the Third International Biogeography Society Conference, Tenerife, January 9–13, 2007.

Ayoub, N.A., Garb, J.E., Hedin, M., Hayashi, C.Y., 2007. Utility of the nuclear protein-coding gene, elongation factor-1 gamma (EF-1 gamma),for spider systematics, emphasizing family level relationships of tarantulas and their kin (Araneae: *Mygalomorphae*). Mol. Phylogenet. Evol. 42, 394–409.

Benjamin, S.P., Dimitrov, D., Gillespie, R.G., Hormiga, G., 2008. Family molecular phylogeny of crab spiders (Araneae: *Thomisidae*). Cladistics 24, 708–722.

Blackledge, T.A., Scharff, N., Coddington, J.A., Sźutz, T., Wenzel, J.W., Hayashi, C.Y., Agnarsson, I., 2009. Reconstructing web evolution and spider diversification in the molecular era. Proc. Natl. Acad. Sci. U S A 106, 5229–5234.

Boll, P.K., 2017. Don't Let Web Bugs Bite. https://earthlingnature.com/2017/01/22/dont-let-the-web-bugs bite/.

Bond, J.E., Opell, B.D., 1998. Testing adaptive radiation and key innovation hypotheses in spiders. Evolution 52, 403–414.

Bosselaers, J., Jocqué, R., 2002. Studies in Corinnidae: cladistic analysis of 38 corinnid and liocranid genera, and transfer of Phrurolithinae. Zool. Scripta 31, 241–270.

Coddington, J.A., 2005. Phylogeny and classification of spiders. In: Ubick, D., Paquin, P., Cushing, P.E., Roth, V.D. (Eds.), Spiders of North America: An Identification Manual. American Arachnological Society, Keene, New Hampshire, pp. 18–24.

Coddington, J., Levi, H.W., 1991. Systematics and evolution of spiders (Araneae). Annu. Rev. Ecol. Systemat. 22, 565–592. https://doi.org/10.1146/annurev.ecolsys.22.1.565.

Darst, C.R., Menendez-Guerrero, P.A., Coloma, L.A., Can-natella, D.C., 2005. Evolution of dietary specialization and chemical defensein poison frogs (*Dendrobatidae*): a comparative analysis. Am. Nat. 165, 56–69.

Elgar, M.A., 1992. Sexual cannibalism in spiders and other invertebrates. In: Elgar, M.A., Crespi, B.J. (Eds.), Cannibalism: Ecology and Evolution Among Diverse Taxa. Oxford Univ. Press, Oxford, U.K, pp. 128–155.

Foelix, R.F., 1996. Biology of Spiders, second ed. Oxford Univ. Press, New York.

Fry, J.D., 1996. The evolution of host specialization: are trade-offs overrated? Am. Nat. 148 (Suppl. l), S84–S107.

Futuyma, D.J., 1986. Evolutionary Biology. Sinauer, Sunderland, MA.

Futuyma, D.J., Moreno, G., 1988. The evolution of ecological specializa- tion. Annu. Rev. Ecol. Systemat. 19, 207–233.

Garwood, R.J., Dunlop, J.A., Selden, P.A., Spencer, A.R.T., Atwood, R.C., Vo, N.T., Drakopoulos, M., 2016. Almost a spider: a 305-million-year-old fossil arachnid and spiderorigins. Proc. R. Soc. B 283, 20160125. https://doi.org/10.1098/rspb.2016.0125.

Gilbert, F., Rotheray, G., Emerson, P., Zafar, R., 1994. The evolution of feeding strategies. In: Eggleton, P., Vane-Wright, R. (Eds.), Phylogenetics and Ecology. Academic Press, London, pp. 323–343.

Grimaldi, D., Engel, M.S., 2005. Evolution of the Insects. Cambridge Univ. Press, Cambridge, U.K.

Harmer, A.M.T., Blackledge, T.A., Madin, J.S., Herberstein, M.E., 2011. High-performance spider webs: integrating biomechanics, ecology and behaviour. J. R. Soc. Interface 8, 457–471. https://doi.org/10.1098/rsif.2010.0454.

Hodek, I., Honek, A., 1996. Ecology of *Coccinellidae*. Kluwer, Dordrecht.

Jermy, T., Ĺabos, E., Molńar, I., 1990. Stenophagy of phytophagousinsects—a result of constraints on the evolution of the nervous sys-tem. In: Maynard Smith, J., Vida, G. (Eds.), Organiza- Tional Constraints on the Dynamics of Evolution. Manchester University Press, Manchester, pp. 157–166.

Jocqúe, R., 1991. A generic revision of the spider family Zodariidae (Araneae). Bull. Am. Mus. Nat. Hist. 201, 1–160.

Jocqué, R., Dippenaar-Schoeman, A.S., 2007. Spider Families of the World. Musée Royal de l'Afrique Central, Tervuren, pp. 336–pp.

Kropf, C., Nentwig, W., Blick, T., Gloor, D., 2019. (Organizing board). (2019 WSC): World spider catalog (version 19.5, Jan 2019). In: Roskov, Y., Ower, G., Orrell, T., Nicolson, D., Bailly, N., Kirk, P.M., Bourgoin, T., DeWalt, R.E., Decock, W., Nieukerken, E. van, Zarucchi, J., Penev, L. (Eds.), Species 2000 & ITIS Catalogue of Life, 2019 Annual Checklist. Species 2000: Naturalis, Leiden, the Netherlands. Digital resource at. www.catalogueoflife.org/annual-checklist/2019.

Joshi, A., Thompson, J.N., 1995. Trade-offs and the evolution of host specialization. Evol. Ecol. 9, 82–92.

Levins, R., Macarthur, R., 1969. An hypothesis to explain the incidence of monophagy. Ecology 50, 910–911.

Li, D., Jackson, R.R., Barrion, A., 1997. Prey preference of *Portia labiata*, *P. africana* and *P. scheltzi*, araneophagic jumping spiders (Araneae: Salticidae). From the Philippines., Sri Lanka, Kenya and Uganda. N. Z. J. Zool. 24, 333–349.

Maddison, W.P., Hedin, M.C., 2003. Jumping spider phylogeny (Araneae: Salticidae). Invertebr. Systemat. 17, 529–549.

Maddison, W.P., Bodner, M.R., Needham, K.M., 2008. Salticid spider phylogeny revisited, with the discovery of a large Australasian clade (Araneae: Salticidae). Zootaxa 1893, 49–64.

Meehan, C.J., Olson, E.J., Reudink, M.W., Kyser, T.K., Curry, R.L., 2009. Herbivory in a spider through exploitation of an ant-plant mutualism. Curr. Biol. 19 (19), R892–R893. https://doi.org/10.1016/j.cub.2009.08.049.

Mitter, C., Farrell, B., Wiegmann, B., 1988. The phylogenetic study of adaptive zones: has phytophagy promoted insect diversification? Am. Nat. 132, 107–128.

Nentwig, W., 1987. The prey of spiders. In: Nentwig, W. (Ed.), Ecophysiology of Spiders. Springer-Verlag, pp. 249–263.

Nosil, P., 2002. Transition rates between specialization and generalization in phytophagous insects. Evolution 56, 1701–1706.

Paradis, E., 2006. Analysis of Phylogenetics and Evolution with R. Springer, New York.

Pekár, S., 2004. Predatory behavior of two European ant-eating spiders (Araneae: *Zodariidae*). J. Arachnol. 32, 31–41.

Pekár, S., Coddington, J.A., Blackledge, T.A., 2011. Evolution of stenophagy in spiders (*Araneae*): evidence based on the comparative analysis of spider diets. Evolution 66 (3), 776–806. https://doi.org/10.1111/j.1558-5646.2011.01471.x.

Pisani, D., Poling, L.L., Lyons-Weiler, M., Hedges, S.B., 2004. The colonization of land by animals: molecular phylogeny and divergence times among arthropods. BMC Biol. 2, 1. https://doi.org/10.1186/1741-7007-2-1.

Platnick, N.I., 2010. The World Spider Catalog, Version 10. American Museum of Natural History. Available at: http://research.amnh.org/entomology/spiders/catalog/index.html. (Accessed 30 June 2009).

Polis, G.A., 1981. The evolution and dynamics of intraspecific predation. Annu. Rev. Ecol. Systemat. 12, 225–251.
Pompozzi, G., Copperi, S., 2018. Feeding frequency, prey acceptance, and natural diet of the mygalomorph spider *Acanthogonatus centralis* Goloboff 1995 (Araneae: *Nemesiidae*). Zool. Stud. 22, 57:e31. https://doi.org/10.6620/ZS.2018.57-31.
Poulin, R., 1992. Determinants of host-specificity in parasites of freshwaterfishes. Int. J. Parasitol. 22, 753–758.
Resende, L., Zepon, T., Bichuette, M., Pape, R., Gil-Santana, H., 2016. Associations between Emesinae heteropterans and spiders in limestone caves of Minas Gerais, Southeastern Brazil. Neotrop. Biol. Conserv. 11 (3). https://doi.org/10.4013/nbc.2016.113.01.
Rypstra, A.L., Samu, F., 2005. Size dependent intraguild predation and cannibalism in coexisting wolf spiders (Araneae, *Lycosidae*). J. Arachnol. 33, 390–397.
Sahney, S., Benton, M.J., Falcon-Lang, H.J., 2010. Rainforest collapse triggered Carboniferous tetrapod diversification in Euramerica. Geology 38 (12), 1079–1082. https://doi.org/10.1130/G31182.1.
Scharff, N., Coddington, J.A., 1997. A phylogenetic analysis of the orb-weaving spider family Araneidae (Arachnida, *Araneae*). Zool. J. Linn. Soc. 120, 355–434.
Schluter, D., Price, T., Mooers, A.O., Ludwig, D., 1997. Likelihood of ancestor states in adaptive radiation. Evolution 51, 1699–1711.
Sebastian, P.A., Peter, K.V., 2009. Spiders of India. Universities Press, p. 614.
Selden, P.A., Shear, W.A., Sutton, M.D., 2008. Fossil evidence for the origin of spider spinnerets, and a proposed arachnid order. Proc Natl Acad Sci U S A 105 (52), 20781–20785. https://doi.org/10.1073/pnas.0809174106.
Shannon, C.E., Weaver, W., 1949. The mathematical theory of communication. University of Illinois Press, Urbana.
Singer, M.S., 2008. Evolutionary ecology of polyphagy. In: Tilmon, K.J. (Ed.), Specialization, Speciation, and Radiation: The Evolutionarybiology of Herbivorous Insects. University of California, Berkley, pp. 29–42.
Stireman, J.O., 2005. The evolution of generalization? Parasitoid flies and the perils of inferring host range evolution from phylogenies. J. Evol. Biol. 18, 325–336.
Toft, C.A., 1995. Evolution of diet specialization in poison-dart frogs (*Dendrobatidae*). Herpetologica 51, 202–216.
Valerio, C.E., 1974. Feeding on eggs by spiderlings of Achaearanea tepidari- orum (Araneae, *Theridiidae*), and the significance of the quiescent instarin ders. J. Arachnol. 2, 57–63.
Vollrath, F., Selden, P., 2007. The role of behavior in the evolution of spiders, silks, and webs. Annu. Rev. Ecol. Evol. Syst. 38, 819–846.
Wygodzinsky, P.W., 1966. A monograph of the *Emesinae* (*Reduviidae, Hemiptera*). Bull. Am. Mus. Nat. Hist. 133, 1–614.
Wignall, A., Taylor, P., 2010. Predatory behaviour of an araneophagic assassin bug. J. Ethol. 28 (3), 437–445. https://doi.org/10.1007/s10164-009-0202-8.
Wise, D.H., Chen, B., 1999. Impact of intraguild predators on survival of a forest-floor wolf spider. Oecologia 121, 129–137.
Yeargan, K.V., 1994. Biology of bolas spiders. Annu. Rev. Entomol. 39, 81–99.
Zhang, S., Chen, H.L., Chen, K.Y., Huang, J.J., Chang, C.C., Piorkowski, D., Liao, C.P., Tso, I.M., 2015. A nocturnal cursorial predator attracts flying prey with a visual lure. Anim. Behav. 102, 119–125. https://doi.org/10.1016/j.anbehav.2014.12.028.

CHAPTER SEVEN

A thumbnail preview of sensuous impression versus sensual expression of the spider

Abstract

Contemporary investigations have shown that the neurophysiological activities of the diverse group known as arachnids are coordinated by the supraesophageal and suboesophageal ganglia, considered the prime neurosensory organs that combine to form the arachnid's primitive central nervous system. Scientists have observed that spiders have a poor sense of direction and poor balance, and as such, their primitive monochromatic vision does not help spiders acting as predators in their attacking sprees or as prey escaping the attacks of predators. However, their acquired neurophysiological traits to intercept and precisely analyze sound waves have helped them attack prey and escape predator attacks. In addition, these traits help them choose their reproductive partners to complete the reproductive cycle, which usually ends in sexual cannibalism.

Keywords: Arachnids; Neurosensory organs; Predator—prey dynamics; Sexual cannibalism.

... males[peacock spiders] use several different aspects to court a female. Males use vibrations early on to gain a female's attention. When they are sure she's watching, they begin to escalate courtship by waving their front legs and showing off their fan. If the female begins ignoring the male, they change their strategy and begin vibrating more.

Michael Kasumovic, Damian Elias, and Madeline Girard

According to conventional knowledge in the literary world, John Milton used the word "sensuous" quite eloquently. First used in 1641, the word describes things related to the physical senses rather than intuition or intellect (Stein, 2015). The ecstatic feeling we get from the fragrance of sandalwood, seemingly similar to the word, supposedly expresses our sensuous feelings or creates a pleasing sensation in our sensory systems. The word "sensual" was reportedly first used in 1425 and was meant to express physical pleasure; through random usage, its meaning has been transformed to the more colloquial sense of sexual pleasure (Borges, 2014).

Spider Evolution
ISBN: 978-0-323-90041-6
https://doi.org/10.1016/B978-0-323-90041-6.00003-6

So sensuality could be further defined as sensuous erotica rather than having a sexual undertone. Before using these two words having a similar tone, one must understand their true meanings to avoid misperceptions. In primitive invertebrate creatures, sense organs were not well developed compared with those of vertebrates, yet the well-dispersed dissemination of arachnids is to some extent related to its use of sense organs. Although a detailed functional elucidation of arachnid sense organs has a limited scope in this discussion, some brief impressions are tenable in understanding their successful dissemination. At the same time, successful completion of the reproductive cycle is quite interesting, as the mating of male and female spiders has gone through a series of biobehavioral incidents and sensual fantasies that must be described here to understand the reproductive system's successful dissemination as well as its primate-dominated world. Often, it's difficult to differentiate these two adjectives of apparently distinct meaning, like sensuous and sensual, when intriguing issues like the origin and evolution of spiders on Earth are concerned.

From anatomical observation, arachnologists have determined that the major part of the cellular structures of the nervous system is within the cephalothorax region, with a clump of nerve cells (called ganglia) in the abdomen (Fabian-Fine et al., 1999; Barth, 2002). Invertebrate neurological scientists consider the nerve tissues "supraesophageal ganglion" as the spider's brain, which is located above the digestive tract and branches out to connect with four pairs of eyes. Scientists have further observed that most sensory interactions and responses of the brain are integrated into the spider's eyes as it coordinates the movements of chelicerae, the release of venom, etc. (Walcott, 1969; Bart, 2002). However, if we start discussing preliminary ideas about the nervous system and sensory organs of any animal, it will begin from the eyes, and the idea that an animal with more eyes obviously has better visual perception is convincing. Because spiders have four pairs of eyes, it seems that they would have better eyesight than humans, as humans have only one pair of eyes. In reality, however, the effectiveness of visual perception does not depend on the number of eyes but rather their structural formation. The anatomical investigation of scientists ascertained that the spider's eyes are very simple and primitive when sense organs are compared with human eyes. All spiders are shortsighted, and they can see neither predator nor prey while sitting a little distance away (Harland and Jackson, 2000). Moreover, they can see properly only when the light reaches their eyes after being reflected from an object from a certain direction, and its vision is monochromatic (it has a close resemblance to the black-and-white scanning type of visual image formation, which is a composite form of an image of light and shade rather than colorful polychromatic vision) (Hsiung et al., 2017).

Just below the digestive system is a lateral connection of the supraesophageal ganglion, a star-shaped body called the suboesophageal ganglion, the branches of which end at legs, palps, etc. This preliminary mode of segmentation has typically also been observed in other arthropods, and it has a striking resemblance to the highly developed human neurological system (Babu, 1969). The segmented distributional mode of suboesophageal ganglial connections in different legs makes it convincible that the sensory impulse coordinates the typical movements of spiders triggered by the blood-streaming of it from the front end to back end through its circulatory system. In recent neurophysiological investigations on the Orb-web spider species *Araneus ventricosus*, scientists opined that spiders' supraesophageal and suboesophageal ganglia are prime neurosensory organs that combine to form the primitive arachnid central nervous system (CNS) (Park and Moon, 2013). However, the critical neurological investigations affirmed that the supraesophageal ganglia are adhered to form a nerve cell cluster, whereas the suboesophageal ganglia comprise the ventral nerve cord and suboesophageal nerve mass of the organized transverse and longitudinal tracts, consisting of peripheral sensory neurons (Gillespie and Spagna, 2009). By understanding the nature and intensity of sound waves and vibration, spiders can determine where predators and prey are located within their environment to decide whether to attack prey or escape from a predator to ensure its survival. Through experimental observation, scientists have ascertained that spiders have a poor sense of direction and poor balance, and therefore, they rely on their primitive visual sense organs, or simple form of eyes, to judge the direction of their movements. A suboesophageal ganglia connection ends in the palp and coordinates the spider's reproductive activities. The unique anatomical findings of arachnologists are that the structural proportion of the suboesophageal ganglion is larger than the supraesophageal ganglion (or brain) in web-building spiders (those are mostly poor in vision) compared with spiders (e.g., salticids) having a better vision (Hsiung et al., 2017). The major impediment to the spider for quick movements to grab prey or escape more quickly from predators is the hydraulic movement executed by its joint legs. To offset the shortcomings of hypoactive visual sense organs and maximize the use of vibration-based sense organs, most spider species have evolved with the formation of an abstract form of art spun by a silk thread, called a spiderweb, that enables the spider not only to entangle its prey but also to take refuge after being chased by predators (Mortimer et al., 2018). The spiderweb has a close resemblance to a biological radar system. To offset their slow movement, spiders use their supplementary sense organs, i.e., the nails at the end of their limbs and hairlike projections on the legs (Wallace, 2017). These nails are often effective in changing the pace of a spider's movement by alterations in touch sensitivity, and these sensitive nails play an important supplementary role in two main sensory actions: sensing the smell of a nearby animal to judge whether it is predator or prey; and if it is prey, to judge by taste whether it is edible.

However, internal and external stimuli, such as visual, touch, thermal, taste, pain, proprioception, biochemical stimuli, cast their impact on the nervous and sensory variations that further witness the changes in physical and physiological functions of spiders in response to the stimuli-induced impetus. All these sensory interactions have broadly been categorized under the following categories:

1. Visual- and temperature-induced sensory interactions

Although most spiders have four pairs of eyes, several spider species have two to three pairs, and some of the tiniest cave-dwelling spiders have no eyes at all. Among those with four pairs of eyes, the large frontal pair of dark-colored eyes are responsible for visual image formation in most spider species (except the species belonging to lycosids and diapsids), scientifically considered anterior median eyes, or AMEs. In contrast, the posterior median eyes, or PMEs, are the most effective image-building eyes for spiders belonging to lycosids and diapsids (Schwab, 2012). Both AMEs and PMEs are known as primary eyes for their respective species, whereas the remaining three pairs of eyes are responsible for surveilling the peripheral zone. They are not very effective in primary image formation but are effective at very low light frequencies. These eyes are regarded as secondary eyes by arachnologists (Infinite Spider, 2014; Morehouse et al., 2017). In terms of structural anatomy, spider eyes have a striking resemblance to human eyes (like a spherical eyeball) in terms of the presence of integral parts such as the cornea and lens. They don't have an adjustable focal length but rather a fixed focal length that renders the spider's eyes mostly ineffective in judging predators or prey from a distance (Morehouse et al., 2017). However, scientists have discovered that by virtue of possessing a structural configuration similar to the telephoto lens of a cup-shaped eyeball, with smaller muscles, salticids are sufficiently good at judging any object from a long distance, which helps this group of spiders to be good hunters and the most capable defenders in the face of predator attacks (Infinite Spider, 2014; Morehouse et al., 2017). The colored iris, which controls the exposition of light to the retinal cup, is absent in the spider's eyeball. Instead, a layer of vitreous pigment cells behind the layer of the cornea controls light traffic. Critical anatomical studies of the spider have exposed the light-sensitive retinal layer from a retinal cup or secondary lens in front of retinal receptor cells, called rhabdomere, to which the optic nerve fibers connect, and it transmits the photo images to the brain (Morehouse et al., 2017).

Scientists have noticed that in the secondary eyes of the spider, special reflective materials, called the tapetum, are observed in the deeper layer of rhabdomere that help the spider see in low light condition or infrared light frequency in nighttime (Infinite Spider, 2014). After a series of investigations, scientists have discovered that the secondary eyes contain the photosensitive pigment, which is functional in the light of 500–540 nm wavelength, which indicates, the secondary eyes could form images effectively in red and blue light (Infinite Spider, 2014; Morehouse et al., 2017). In contrast, the primary eyes can form an image in the ray of light having a wavelength around 360 nm, which is turquoise in color, almost ultraviolet light. Many other photoreceptors have been discovered in the eyes of the spider that can form images in various colors and light ranges: turquoise, orange, near-ultraviolet, blue-violet, etc. But such findings have been challenged by an opposing group of scientists (Infinite Spider, 2014; Morehouse et al., 2017). In an experimental observation on neurological functions in response to visual stimuli on the Jumping spider, genus *Phidippus* (Salticidae), supraesophageal ganglia of the CNS mediate all visual sensory impulse and image formation processes and coordinate the physical, physiological, and behavioral actions in response to such visual stimuli by maintaining the logistics through the optic nerve connectors between primary and secondary eyes and CNS (Hill, 2006).

2. Vibration- and touch-induced sensory interactions

To overcome its limited visual potential, the spider relies on using the alternative sense organs that are vibrations and sound-mediated touch-based sensuous receptions to respond to any impetus posed by external or internal stimuli (Barth, 2002). Arachnologists observed that the entire surface of the body and the tip of the limbs are covered with different shapes, sizes, and thicknesses of the hairlike sinuous structures responsible for touch sensitivity and reception of vibrations (Walcott, 1969; Barth, 2002; Gillespie and Spagna, 2009). Several spiders have long, slender hairs on their legs called "trichobothria," and the "lyriform organ," a ten-slit structure adhered to the tarsal and metatarsal parts of the legs, is responsible for receiving low-frequency sound waves and airborne vibrations (Walcott, 1969; Santer and Hebets, 2011). As primitive invertebrates, spiders are devoid of the ears that vertebrates use to receive and analyze various sound waves and instead rely on the trichobothria and lyriform organ, the primitive organs capable of vibration and that can receive sound waves with frequencies as high as 1400 Hz frequency, analogous to the earlike organs of vertebrates for nervous and sensory communicative interactions with other animals (Barth, 2002).

3. Biochemical and thermal sensory interactions

The tiny cavity in the tarsus region of the spider's legs is called the tarsal organ, and scientists consider this organ a key receptor of pheromones (Barth, 2002). The tarsal organ is believed to be involved in adjudging humidity and hot and cold situations. In terms of sense of smell, scientists have noticed that the smelling sense of spiders is absolutely amazing, as spiders have evolved with a fascinating biochemical sensory system, where female spiders appear in a nonmotile state but try to communicate with male spiders for mating, during which they emit stinky biochemical defense odors called camouflage with a smell that attracts their male counterparts (Langley, 2013). Arachnologists have noticed that some spider species emit stinky odors or camouflage to pass a message to other community members that another species occupies the habitat they want to settle in. Hence, arachnologists have further interpolated that sense of smell is an important mode of communication used by spiders in the interspecific struggle for existence. Male candidates are a little smaller than their female counterparts and are more proactive in hanging around and tracking down the source of smell (the pheromones) using their sensory receptors to reach their mating partners (Barth, 2002). The pheromones are of low molecular weight, lipid-soluble, volatile chemicals, and it has a close biochemical resemblance to the aromatic terpenoids of aromatic plants. Hence, this sensuous communication employing pheromone secretion to communicate for courtship or mating has been recognized as sensual communication.

Most spiders have evolved with a molecular thermosensor, which motivated them to move from moist and cold places to dry and hot places to weave their webs, though several spider species occupy ecological niches in Arctic and Circumpolar regions. Hence, the species diversity of arachnids is at its maximum in tropical and subtropical regions rather than temperate and polar regions (Barth, 2002). However, scientists have tried to put the fragmentary observations on the sensory and nervous system of spiders, as no intensive research has not been undertaken in an attempt to unearth the structural and functional model of the spider's CNS (Japyassu and Laland, 2017). The colloquial perception is that spiders are fearsome creatures, and they have earned a bad reputation that they often attack humans or other animals. That's not actually true, as the spider is a vulnerable animal often threatened by survival challenges imposed on it by anthropogenic activities and other predators (Barth, 2002). Normally, scientists have observed that any provocative attack, accident, or survival challenge is typically transmitted through spiders' sensory and nervous systems, which turn on defense systems that provoke them to attack humans or annoy predators who jeopardize their survival or existence (Japyassu and Laland, 2017).

4. Proprioception—the ability to sense stimuli evolving inside the body in terms of position, movement, and equilibrium

Spiders use proprioceptors to determine the posture of their body appendages, such as the orientation of leg segments relative to the orientation of chelicerae, cephalothorax, abdomen, etc., and they maintain an equilibrium status in any position—horizontal, upside down, angular—when entangled in their web (Barth, 2002). To maintain an equilibrium state when running, jumping, or in any other position that represents a drift from a stable position, humans typically use a wide range of receptors (visual, audible, etc.), whereas spiders are thought to use legs and leg joint proprioceptors to maintain their dynamic equilibrium. The proprioceptors used by spiders are integrated into their leg joints (Seyfarth et al., 1985). The neurological investigations and anatomical findings indicate that spiders belonging to the ctenid group possess sensory hair plates known as "coxal hair plates," and internal capsule receptors at the leg joint are responsible for neurological monitoring of flexion or extension at the leg joints at particular points in time (Seyfarth et al., 1985). Apart from sensory hairs, the lyriform slits at the metatarsus consist of a layer of tissues, and slits between act as a key proprioceptor in responses to tension (Schaber and Barth, 2015). Other types of tiny slits in the palp, pedicels, and spinneret joints act as minor proprioceptors responsible for adjudging the orientation of the abdomen and cephalothorax, dealing with impact- or deformation-related issues, transmitting sensory impulses to the brain, and carrying out nervous impulses to coordinate the next course of action, such as fleeing from predators or attacking prey (Schaber and Barth, 2015). Scientists have noticed that proprioceptors play a key role in mating, whereby some females enjoy the right to choose their male counterparts. Contemporary investigations on the leg proprioceptors of female *Argiope keyserlingi* spiders show that the females adjudge the gravity and vibrations caused by their male counterparts and select an eligible mating partner for courtship (Herberstein et al., 2002). Many arachnologists consider this a typical mode of courtship within arachnids, known as "crazy courtship" (Tomas, 2008), as it involves the whole range of crazy actions: psychedelic rituals, genital mutilation, orifice plugging, sexual cannibalism, etc. Scientific studies have confirmed that for the Redback spider (*Latrodectus hasselti*), it takes 4–5 h to complete this courtship ritual. When a noncompeting single male proceeds one-to-one to become involved in courtship with a female counterpart, the duration is much shorter (typically 40 min to 1 h) than when several males compete to win over a female for mating (Tomas, 2008).

In general, arachnologists have observed that female spiders emit pheromones that trigger a social responsibility in their male counterparts within the same species. In responding to such chemical factors and sex pheromones, male spiders begin a search for a female to mate. To make themselves safer, several males typically look for newly molted females to mate, as they are physically too weak to kill the males during mating. The mating of spiders is a cumbersome process characterized by the sexual cannibalism of female spiders. Most male spiders across the diverse group of species possess a couple of mating organs in their lower abdomen called pedipalps, and one is used with each act of copulation, so theoretically, the male spider has two possible chances of copulation in their lifetime unless their mating partner devours them. Experimental observations on male dark fishing spiders (*Dolomedes tenebrosus*) revealed that the male genitalia become disfigured at the end of copulation, which causes their death, and the female feeds on the dead carcass of its mating partner (Lewis, 2013). Often the male spider strives for an elaborate courtship to save itself from being eaten by its female mating partner. Scientists have observed that sometimes females don't like their potential mating partners, in which case they devour them before mating, but most often, female spiders devour their partners after mating. Many times, the female spiders just let their male partner go unharmed after mating, and scientists have critically observed a fair chance of survival for males that finish the act of copulation in 10 s or less. Experimental observation reveals further that the males that mated with a sibling took less time for copulation, so they have better survival opportunities than those male counterparts that partnered with nonsiblings and took longer to finish copulation, and thus became the prey of sexual cannibalism (Lewis, 2013). It is difficult to predict in advance what would be the possible scenario when a couple of spiders engage for courtship or engage in mating. Scientists have observed that male spiders are typically much smaller than their female counterparts, so they are not bold enough to make them calm and less likely to control their aggressive behavior, but if they did, the female spiders would become less aggressive and abstain from harming their male mates.

However, the male spiders start vibrating the webs to signal female spiders for mating. In the case of jumping spiders, male spiders have been seen shaking the glowing patches of white and green colored skin of their body, which creates a glowing reflection in UV light, and they give a visual signal by initiating a type of shivering posture (Tomas, 2008). At the same time, the female counterparts of the jumping spiders respond with a bright glowing hue in UV light. These signals indicate their interest in the copulation process (Tomas, 2008). To convince their intended communication for copulation, in some species, the males start moving around the female counterparts for courtship, and that particular movement of male spiders is recognized as courtship dancing of spiders. It depends on the nature of the aggressive mode of behavior; if females are not interested in responding to the courtship signals of the male counterpart in their habitat, the females tend to kill and eat them or simply leave the area. It's difficult to predict whether a female will participate in the courtship ritual with a particular male. But it has often been observed that most females become hyperactive and aggressive during or before mating.

An interesting experimental observation of Steven Schwartz, an eminent researcher in arachnid behavioral ecology at the University of Nebraska–Lincoln, USA, was that male fishing spiders engaged in mating with its female counterpart just once in their lifetime (their this characteristic is known as monogyny in the animal world) and they possibly die out of exhaustion or from any internal factor (Lewis, 2013). From his critical observation, Steven Schwartz inferred that unlike a living male being devoured by a female at the end of mating, the pedipalps inflates at ejaculation, and its sudden deflation after that renders the male motionless, and then he dies (Lewis, 2013). So Schwartz disagreed with the active "sexual cannibalism" hypothesis and rather designated this course of action hypothetically as "monogyny and self-sacrifice," which has also been observed among Australian redback spiders (Lewis, 2013).

Scientists have observed Another crazy courtship ritual in the case of a spider species named *Tidarren argo*, where male spiders engage in self-mutilation of one of their palps for better mobility (Michalik et al., 2010). They generally lose their remaining palp during copulation, as it becomes stuck in the female genitalia or epigyne and is torn off due to the female spider's aggressive behavior during the end phase of copulation. The severed male genitalia is stuck in the epigyne for about 4 h (Tomas, 2008). So at the end of the courtship, the male partner dies. That is why females of the *T. argo* species try to choose another male partner to fertilize their eggs. So two males participate in the successful completion of the reproductive cycle of female *T. argo* spiders. Another weird sexual practice has been observed in the Orb-web spider species of *Argiope bruennichi*, where the tip of the adult male genitalia is torn off during first-time copulation to permanently block the female's sexual orifice from future mating efforts, and scientists have critically observed that second and subsequent copulation efforts by *A. bruennichi* females remain unsuccessful, and copulation lasts for less than 8 s (Tomas, 2008).

There is a typical arachnid courtship practice witnessed among the species *Physocyclus globosus* that is called "stridulations," in which the female spider emits high-pitched squeaks while mating, typically in response to the squeezing actions of the male genitalia inside the female orifice during sexual intercourse (Tomas, 2008). The squeaks from the female act as an expression or communication to the male spider that his genitalia are squeezing too hard, and the prolonged intercourse has gone on for too long. But this shrill cry has dual meaning to male and female spiders engaged in copulation. The male spider considers it a sign of a winning manhood over competitors and attempts to maximize its chance of being selected over competing males by the potential mating partner. At the same time, such squeaking by the female prepares her to end the lovemaking process and kill her partner immediately after copulation and then eat him. Many arachnologists believe that females cannot suppress their indomitable hunger during copulation, which leads them to kill and feed on their mates (which are physically smaller and more easily available in their environment) at the end of copulation. On most occasions, whenever the practice of sexual cannibalism takes place, the female spiders are much larger than their male counterparts. Arachnologists refer to wolf spiders (*Hogna helluo*) here to substantiate that about 80% of the time, smaller-sized males are subjected to the hunger of female spiders during sexual cannibalism, though the story is different in the case of blood-drinking Jumping spiders (*Evarcha culicivora*), in which larger-sized males, rather than smaller-sized females, are cannibalistic after a female spider has been deflowered by a male (Tomas, 2008). Scientists have observed that in response to second and subsequent efforts by a large male to mate with the same female after it has lost its virginity, the female looks for a smaller-sized male mating partner to minimize its chance of being eaten by the larger-sized male at the end of the love-making process.

There is a unique courtship strategy adopted by some male spiders (affecting around 30 species) in which they tie the front legs of the females with silk cords before engaging in copulation. Scientists have observed that the nursery web spider (*Pisaurina mira*) engages in spinning silks in the air for this courtship ritual, known as "bridal veils" (Anderson and Hebets, 2016). Arachnologists have observed that this unique strategy employed by the Nursery web spider effectively warrants two important issues. As the females are much larger than their male counterparts, there is a threat for the male spider that its female mating partner will eat it sometime during the courtship process, so tying the front legs ensures the male's physical safety (Anderson and Hebets, 2016). The second factor is quite interesting from a biobehavioral and evolutionary standpoint when considered from the perspective of sexual selection. It has often been observed that when female spiders engage in sexual intercourse with more than one male partner (polyandrous), it does not necessarily mean that the female wants its egg to be inseminated while engaged in copulation of each male partner. The female has the discretion to pick any male partner to inseminate her and fertilize her eggs, and this intensive sexual selection process generates a sharp surge in population and allows spiders to occupy nearly all the diverse ecological niches of the world. However, the ritual of tying the front legs with silk cords helps males extend the duration of copulation, letting them ensure the optimum potential of insemination of their love mates as well as maximizing the opportunity of fertilization of eggs. In a 2016 interview, Catherine Scott, the famous arachnologist from the University of Toronto, Canada, revealed her critical investigations that males of the nursery web spiders engaged in spinning the bridal veil have added a "cocktail of pheromones" to lure the females to become more intensively involved in the lovemaking process (Greshko, 2017). However, the arachnids' typical mode of sexual cannibalism by female spiders during the reproductive cycle is an unprecedented course of action witnessed in the animal kingdom, and the reason for such biobehavioral action is not clear to scientists, though several hypotheses have been promulgated to define it. For animals generally, successful copulation ends with a fusion of egg and sperm, but the case of spiders is unique in that no direct fusion of egg and sperm takes place. Arachnologists have described the process as follows (Tomas, 2008): "Males ejaculate onto ready-made small sperm webs and then transfer their sperm to syringe-like structures on the tips of their front appendages, or palps. As the courtship progresses for the male Jumping spider, he will arch his body, vibrate his palps and slink on tiptoe toward the female. If courtship is successful, the male injects his sperm from his palps into the female's genital opening, known as the epigyne, on the underside of her abdomen." Reproductive biologists and arachnologists have noticed that the male spiders often discharge sperm somewhere on the dragline of the web, and it is somehow transferred to the tip of the female genitalia in the course of her movement on the web.

The passive fusion of sperm and eggs leads to the formation of eggs, and the female spider often lays the eggs and then dies, or she carries the eggs in the egg sac until the newborns emerge, and she dies immediately thereafter to complete the process. It's not clear to reproductive biologists whether this egg-laying process may be so energy-draining for female spiders that they succumb to death. The sudden biobehavioral change in the female spider's nature that renders her hyperactive, and the death of male spiders during the copulatory or postcopulatory phase, and the death of the female spider herself after laying eggs are not well-defined events in the reproductive cycle of spiders. So there is a myth about juvenile spiders that newborn spiders have never been nurtured by their mothers like other animals, but there are some exceptions, too, where the juveniles have been cared for by their mothers. So far, scientists don't have much information about the bonding process between a mother and a juvenile. It seems that the shortened life spans or sudden deaths of male and female spiders during the mating and egg-laying phase might have some relation with successful occurrence of the reproductive cycle. Perhaps this complex process of arachnid reproduction cycle, which is not very convincible to scientists, leads to the birth of a maximum number of offsprings at the cost of the life of male and female adults. From a sociobiobehavioral viewpoint, it seems to be an animal's instinct to meet its biological needs to maximize its presence in maximum strata of ecological niches by a surging mode of multiplication rather than spending time to develop a bonding process with juveniles and nurturing them afterward.

Although the number varies from species to species, it has been observed that female spiders generally lay about 3000 eggs that they carry in their egg sacs until the eggs are ready to be hatched. As postnatal care of the newborn spiders is missing, a substantially higher rate of juvenile mortality has been witnessed among spiders, and that seems to be a primary reason that spiders lay such an extraordinary number of eggs to stay in the evolutionary race of "struggle for existence". Even after passing the juvenile phase, spiders are subjected to stiff survival challenges in terms of intergeneric threats (like predatory attacks), interspecific struggles (like the threat from araneophagic species) to a large extent, and intraspecific competition (the struggle for food resources and space) that stunt young adults in attaining mature adulthood and causing them to die before attaining reproductive age. Scientific observations reveal that in wild habitats, most spiders live for 3–4 years, whereas they have been found to live longer under captive conditions. Tarantulas are a perfect example of this—their lives are extended by 15–20 years under captive conditions (Meixler, 2018). The main reason for such a twofold to threefold extension of life span under captive conditions relative to a member of the same species in a wild habitat is thought to be the result of a high frequency of mating in wild conditions, whereas spiders in captivity have less chance of joining the reproductive cycle. Even though we have observed the series of weird courtship rituals of spiders, we cannot ignore that since their evolution on Earth as inconsequent nonvertebrate creatures, spiders have engaged in sexual selection for more than a million years. This gives the impression that the widespread dispersal of this bizarre animal competing in a "high-stakes competition" to choose a mating partner makes it a picky lover within the animal kingdom. The spider has managed to survive for a long time despite numerous survival challenges, and it has ensured its sustainability to this point even in the most favored regimes of vertebrates (Cory and Schneider, 2017).

References

Anderson, A.G., Hebets, E., 2016. Benefits of body size dimorphism and copulatory silk wrapping in the sexually cannibalistic nursery web spider, *Pisaurina mira*. Biol. Lett. 12, 20150957. https://doi.org/10.1098/rsbl.2015.0957.

Babu, K.S., 1969. Certain histological and anatomical features of the central nervous system of a large Indian spider, Poecilotheria. Am. Zool. 9, 113–119.

Barth, F.G., 2002. A Spider's World: Senses and Behavior. Springer, Berlin, p. 394.

Borges, M., 2014. Passions and evil Kant's philosophy. Manuscrito 37 (2), 333–355. https://doi.org/10.1590/S0100-60452015005000003. Epub January 13, 2015.

Cory, A.L., Schneider, J.M., 2017. Effects of social information on life history and mating tactics of males in the orb-web spider Argiope bruennichi. Ecol. Evol. 8 (1), 344–355. https://doi.org/10.1002/ece3.3672.

Fabian-Fine, R., Hoger, U., Seyfarth, E.-A., Meinertzhagen, I.A., 1999. Peripheral synapses at identified mechanosensory neurons in spiders: three-dimensional reconstruction and GABA immunocytochemistry. J. Neurosci. 19 (1), 298–310. https://doi.org/10.1523/JNEUROSCI.19-01-00298.1999.

Gillespie, R.G., Spagna, J.C., 2009. Spiders. In: Resh, V.H., Carde, R.T. (Eds.), Encyclopedia in Insects, second ed. Elsevier, Londin, pp. 941–951.

Greshko, M., 2017. Bondage, Cannibalism, and Castration— Spiders' Wild Sex Lives. www.nationalgeographic.com/news/2017/02/spiders-sex-valentines-day-cannibalism/.

Harland, D., Jackson, R., 2000. Eight-legged cats' and how they see - a review of recent research on jumping spiders (araneae: salticidae). CIMBEBASIA 16, 231–240.

Herberstein, M., Schneider, J., Elgar, M., 2002. Costs of courtship and mating in a sexually cannibalistic orb-web spider: female mating strategies and their consequences for males. Behav. Ecol. Sociobiol. 51, 440–446. https://doi.org/10.1007/s00265-002-0460-8.

Hill, D.E., 2006. The Structure of the Central Nervous System of Jumping Spiders of the Genus *Phidippus* (Araneae: Salticidae). In: Central Nervous System of Phidippus [RV1], pp. 1–46.

Hsiung, B.-K., Justyn, N.M., Blackledge, T.A., Shawkey, M.D., 2017. Spiders have rich pigmentary and structural colour palettes. J. Exp. Biol. 220, 1975–1983. https://doi.org/10.1242/jeb.156083.

Infinite Spider, 2014. How Many Eyes Does a Spider Have? https://infinitespider.com/many-eyes-spider/.

Japyassú, H.F., Laland, K.N., 2017. Extended spider cognition. Anim. Cogn. 20, 375–395. https://doi.org/10.1007/s10071-017-1069-7.

Langley, L., 2013. 5 Animals with Stinky Defenses. https://www.nationalgeographic.com/news/2013/9/animals-defenses-smells-predators/.

Lewis, T., 2013. Tough Love: Male Spiders Die for Sex. https://www.livescience.com/37536-spiders-die-for-sex.html.

Meixler, E., 2018. The World's Oldest Known Spider Has Died at the Age of 43 from a Wasp Sting. https://time.com/5259382/worlds-oldest-known-spider-dies-43/.

Michalik, P., Knoflach, B., Thaler, K., Alberti, G., 2010. Live for the moment – adaptations in the male genital system of a sexually cannibalistic spider (theridiidae, araneae). Tissue Cell 42, 32–36. https://doi.org/10.1016/j.tice.2009.06.004.

Morehouse, N., Buschbeck, E., Zurek, D., Steck, M., Porter, M., 2017. Molecular evolution of spider vision: new opportunities, familiar players. Biol. Bull. 233. https://doi.org/10.1086/693977.

Mortimer, B., Soler, A., Siviour, C.R., Vollrath, F., 2018. Remote monitoring of vibrational information in spider webs. Sci. Nat. 105, 37. https://doi.org/10.1007/s00114-018-1561-1.

Park, Y.-K., Moon, M.-J., 2013. Microstructural organization of the central nervous system in the orb-web spider Araneus ventricosus (araneae: araneidae). Appl. Microsc. 43. https://doi.org/10.9729/AM.2013.43.2.65.

Santer, R., Hebets, E., 2011. The sensory and behavioural biology of whip spiders (arachnida, amblypygi). Adv. Insect Physiol. 41, 1–64. https://doi.org/10.1016/B978-0-12-415919-8.00001-X.

Schaber, C.F., Barth, F.G., 2015. Spider joint hair sensilla: adaptation to proprioreceptive stimulation. J. Comp. Physiol. A Neuroethol. Sens. Neural. Behav. Physiol. 201 (2), 235–248. https://doi.org/10.1007/s00359-014-0965-4.

Schwab, I.R., 2012. Evolution's Witness: How Eyes Evolved. Oxford University Press, p. 328.
Seyfarth, E., Eckweiler, W., Hammer, K., 1985. Proprioceptors and sensory nerves in the legs of a spider, Cupiennius salei (arachnida, araneida). Zoomorphology 105, 190–196.
Stein, S., 2015. Ornate Rhetorick. The Paris Review. https://www.theparisreview.org/blog/2015/07/02/ornate-rhetorick/.
Tomas, A., 2008. Very Strange: The Spider Sex Chronicles. https://www.livescience.com/5233-strange-spider-sex-chronicles.html.
Walcott, C.A., 1969. Spider's vibration receptor: its anatomy and physiology. Am. Zool. 9 (1), 133–144. https://doi.org/10.1093/icb/9.1.133.
Wallace, T., 2017. The Science of Spider Paws. https://cosmosmagazine.com/biology/the-science-of-spider-paws/.

CHAPTER EIGHT

Untangling the mystery of the origin and evolution of spiders on planet Earth

Abstract

The origin and evolutionary footsteps of arachnids are nearly untraceable in the arachnid genome. Evolutionary geneticists and molecular biologists have contemplated that the DNA-encrypted molecular footprints of the evolutionary progress of arachnid life have likely been tarnished. Furthermore, scientists have witnessed that the prehistoric evolutionary history of arachnids has gone through a complex, reticulate trail of diversification wherein a series of mutations has irreversibly tarnished the stepping-stones required to join the dotted line to make the sketch complete from beginning to end. Computed tomographic analysis to perform molecular backtracking of the oldest paleontological evidence discovered at the fossil site of Upper Carboniferous in France appears to indicate that primitive land dwellers transmigrated from water bodies and evolved around 420 million to 300 million years ago.

Keywords: Arachnids; Evolution; Paleontological evidence; Upper carboniferous.

Any sort of flagelliform appendage tends to be like an antenna, It's for sensing the environment. Animals that have a long whippy tail tend to have it for sensory purposes. The ones we recognized previously were different in that they had a tail but don't have the spinnerets. That's why the new one is really interesting, apart from the fact that it's much younger—it seems to be an intermediate form. It makes us wonder if these may still be alive today.

Paul Selden

"It's not quite a spider, but it's close to being one," Russel Garwood, the popular paleontologist at the University of Manchester in the United Kingdom, claimed unhesitatingly in an interview session of "Live Science," referring to his meticulous study of a fossil of "almost spider" (which has been scientifically recognized as *Idmonarachne brassierei*) retrieved from an iron carbonate ore at a 305 million-year-old fossil site of siderite concretion in Eastern France (Pappas, 2016). Garwood, considered a connoisseur in the field of evolutionary biology, further commented that arachnids, one of the

Spider Evolution
ISBN: 978-0-323-90041-6
https://doi.org/10.1016/B978-0-323-90041-6.00004-8

primitive groups of eight-legged animals, comprised scorpions, ticks, and spiders. The evolutionary history of arachnids has not emerged from the overcast shadow of an evolutionary labyrinth of primitive invertebrates that supposedly gave rise to the origin and evolution of contemporary spiders in the course of stepwise evolutionary strides of life (Garwood et al., 2016). In support of his careful and explicit crafting of the terminology, "almost spider," Garwood has further substantiated the lack of well-preserved rock samples at later fossil sites, so ideas about primitive life, specifically the invertebrates that existed during that period, are only poor assumptions, and the possibility of obtaining paleontological evidence about ancient arachnids is limited. In response to a question of the feasibility of alternative appraisal of origin and evolutionary treatise of spiders by decrypting molecular footprints written in its DNA using technological advances in molecular genetics and genomics, Garwood responded that it's very difficult to decipher such molecular footprints on the evolution of arachnids. Furthermore, he opined that the sequence of molecular footprints encrypted in arachnid DNA has likely been tarnished, as the prehistoric evolutionary history of arachnids has gone through a reticulate trail of diversification whereby a series of mutations has supposedly burnt the bridge of the evolutionary trail in several locations and rendered the evolutionary history of arachnids untraceable in the arachnid genome (Pappas, 2016).

Although the oldest paleontological evidence of an arachnid creature (10 mm long) was discovered decades ago, there were many difficulties in studying it, as its front was completely buried in the ore of iron carbonate rocks of siderite concretion. However, the latest computed tomographic studies (synchrotron-based phase-contrast computed tomography reconstruction) of the "almost spider" fossil retrieved from the Montceau-Les-Mines, the subterranean coal reserve in Eastern France, have provided both a morphological analysis and the structural formation of the buried legs and mouthparts of this prehistoric creature. Along with taxonomic analysis at the generic and species levels of this arachnid, Garwood and his colleagues successfully determined the geological age from Upper Carboniferous of this ancient creature, which was 299–305 million years old, and contemplated to be evolved as the primitive group of land dwellers, supposedly adopting a terrestrial life, after migrating from water bodies around 420 million years ago (Garwood et al., 2016). Evolutionary investigations by Garwood et al. (2016) on this oldest, fossiliferous upper Paleozoic spider specimen further revealed that this creature shares an ecological niche with its "nonspider phylogenetic cousins" of the Paleozoic arachnid grades. However, in paleontological investigations related to the earliest fossil of a primitive spiderlike creature known as *Attercopus fimbriunguis*, a group of arachnologists hypothesized that around 400 million years ago (mya), the thick-waisted, spiderlike creatures evolved as prehistoric arachnid ancestors. Meanwhile, the thin-waisted true spiders, with abdominal segmentation and web-spinning organs like spinnerets, diverged from their prehistoric arachnid progenitors during 380 mya in Devonian, which has been taken place around 150 mya before when the dinosaurs appeared on this Earth (Selden et al., 2009).

The critical anatomical investigations of the legs and the mouthparts of the "almost spider" creature, supplemented by paleontological and evolutionary analyses done by Garwood and his associates have helped scientists affirm their view about designating it the "almost spider," as this unique creature had no spinnerets and a segmented abdomen (rather than a fused abdomen), the typical features of a true spider (Pappas, 2016). Such observations have led team Garwood to hypothesize the "almost spider" as a phylogenetic cousin of the modern spiders, as he said, "We're looking at a line of spider-like arachnids that haven't survived but must have split off before 305 mya." (Pappas, 2016). Garwood et al. (2016) further observed that the primitive arachnid member of Uraraneida from the Devonian—Permian era (385—275 mya) is phylogenetically represented by the extinct 305 million-year-old fossil taxon *Idmonarachne brasieri* or "almost spider," with a taillike structure (called flagellum) but no spinnerets. Although the members of Uraraneida did not yet have any spinnerets, they possessed spigots that could produce silk threads. It should be mentioned that spiders that evolved around 300 mya are characterized by the presence of spinnerets, considered "modified legs" by arachnologists, that exude silk through a process controlled by the spigots. The male spider possesses another "modified leg" between their four pairs of legs and fangs, which is typically used to insert sperm into the female spider at the time of mating (Pennisi, 2018). However, Garwood and his fellow scientists promulgated that *I. brasieri* might have produced the silk but might not have articulated the weaving potential of the silk web because of the absence of structures like spinnerets (Garwood et al., 2016). Going one step further, Garwood and his colleagues engaged in cladistic analysis to reconstruct the phylogenetic tree of arachnids to obtain a plausible interpretation of the origin and evolution of spider, but it was a cumbersome process to accommodate spiders, scorpions, and opiliones in the same arachnid phylogenetic tree, which prompted Garwood to say, "*Arachnids as a whole are an incredibly successful group ... They're the most diverse group of living organisms after insects ..... They're successful—but we have a very limited understanding of how they are related to each other*" (Garwood et al., 2016). Based on composite studies of paleontological explorations and cladistic analysis, evolutionary biologists and the arachnologists have contemplated further that prehistoric spiders used to have a "smooth back" rather than the segmented abdomens characterizing scorpions, and the spiders supposedly diverged from the prehistoric arachnid mother-stock around 430 mya (Pennisi, 2018).

According to paleontological investigations and analysis, most prehistoric fossil spiders (extinct) with abdominal segmentation belonged to the mesothelial that used to possess spinnerets under the middle part of the abdominal segment, whereas the spinnerets of modern-day (extant) spiders are found at the end of the abdominal segment. The findings of several fossilized insects like cockroaches, silverfish, and millipedes from the same fossil site as that of the fossil spiders in the Paleozoic Age led scientists to hypothesize that primitive spiders may have acted as predators to hunt a wide range of insects as prey (Pappas, 2013). It has further been assumed that the silks secreted by the spinnerets of primitive spiders were most likely used for protecting its eggs (as a protecting cover for the eggs) or making a lining for a retreat hole or constructing a trapdoor for a protective role rather than as a hunting device. As far as a fossil representative of arachnids is concerned, the 300 million-year-old fossil arachnid *Megaarachne servinei* is considered one of the largest extinct spider species, with a half-meter-long body combined with a shovellike jaw and a protective shieldlike covering on the abdomen (Selden et al., 2005). It seems that *M. servinei* might have a wide array of prey on its platter. But the evolution and existence of large-sized spiders like arachnids or spider progenitors with massive abdominal shields indicate the presence of larger-sized predators of these giant, fossilized spiderlike arachnids (Switek, 2010). However, with time, there was a big change in the diversification of plant and animal life that happened in the biosphere, and against the backdrop of such paleoclimatic changes along with drastic changes in paleoflora and paleofauna around 250 mya, spinnerets evolved with the abdominal segment settled at the end, and arachnologists recognized these newly evolved spiders as modern-day spiders, belonging to Opisthothelae, that spin sheet-webs and web mazes to develop safety draglines to capture prey on the ground and in foliage (Pappas, 2013). From observation and interpretation of the paleontological data available for the Jurassic Age (around 191–136 mya) when the dinosaur roamed the world, scientists have promulgated that there was a surge of flying insects on Earth, indicating that the use of specialized web-spinning potential in the members of Opisthothelae evolved as a coevolutionary process to maintain predator–prey dynamics in an equilibrium state (Goulson, 2014). All these early fossil specimens of spiders and spiderlike creatures would inevitably give an idea of the origin and evolution of spiderlike arachnids, but the availability of the amber-trapped true spider fossils from the tropical rainforest in South and Southeast Asia, which are 30 million years old, provided a clear indication that the evolution and diversification of modern-day spiders happened in the Tertiary Period against the backdrop of a surge in flowering plants and angiosperms on Earth (Pappas, 2013). According to the contemporary findings of arachnologists, the prehistoric segmented members of Mesothelin are often found in an extant state in the tropical forest ecosystems of Eastern Asia (particularly in the dense forests of China and Indonesia), and phylogenetic investigations ascertained further that the extant ancestral lineages of prehistoric spiders that evolved around the Paleozoic Period are structurally unchanged, so these prehistoric extant lineages of Mesothelid spiders should be recognized as "living fossils" (Xu et al., 2015).

On the other side, the evolutionary story of spiders has become an integral part of amber mining and its commercial use in Myanmar, where the old, naturally preserved amber has been widely used in the jewelry industry for years. Eventually, the well-preserved prehistoric ambers trapped many invertebrates and turned them into well-preserved fossil specimens that have drawn the attention of evolutionary biologists, specifically scientists, engaged in unfurling the origin and evolution of the prehistoric invertebrate creatures for ages. At the beginning of 2018, Prashant Sharma, an evolutionary developmental biologist at the University of Wisconsin at Madison, USA, and his associates were engaged in investigating four specimens of such fossiliferous, amber-entrapped prehistoric arachnid members, which were assumed to be 100 million years old (Pennisi, 2018). The critical investigation of these amber-encased prehistoric arachnid specimens from Myanmar done by Sharma and his colleagues ascertained that they looked like an intermediate organism between spider and scorpion, and based on his careful observations, Sharma said that this "could help close major gaps in our understanding of spider evolution" (Penniai, 5th February 2018). The beginning of the story to highlight the evolutionary journey of the spider started with the discovery of a spigot-bearing fossil of an arachnid specimen, and age-dating confirmed that it was around 100 million years older than the earliest known spider (Pennisi, 2018). After a period of paucity, this issue resurfaced in 2008 when paleontologists tried to place the paleontological evidence with contemporary findings on molecular genetics and genomics and hypothesized that the prehistoric biological "silk producers" were not necessarily true spiders, but rather, they would be better treated as the evolutionary predecessors of modern-day spiders, the arachnid stepping stones. Such findings have led arachnologists to place such "evolutionary stepping stones" in the arachnid group Uraraneida that evolved 450–250 mya (Pennisi, 2018). Although arachnologists and evolutionary biologists have put forward the "spigot-carrier model" to narrate the evolutionary journey of the modern spiders, this story has left one important issue to address, as scientists did not mention or narrate anything about the transitional changes in arachnid members that supposedly shifted from spigot bearers to spinneret bearers: How did it happen and when did it happen? And so forth.

Recently, two eminent paleontologists at Nanjing Institute of Geology and Paleontology, China, discovered amber-trapped fossil specimens of prehistoric arachnids (which were approximately 5 mm long and belonging to Uraraneida) collected from Myanmar to Wang Bo and Huang Diying. Wang Bo and his team examined the fossil specimens and concluded that the fossil specimen was an extinct arachnid belonging to uraraneids and considered it the ancestor of the modern spider. Wang Bo published his work on this fossil specimen as a new species with the name *Chimerachne yingi* (which means a chimeric spider in Latin) (Pennisi, 2018). At the same time, Huang Diying and his team published their work in the same journal (*Nature Ecology and Evolution*) after completing their investigations, and surprisingly, the major findings of both investigations are more or less the same with some minor differences. Greg Edgecombe, the paleontologist at the Natural History Museum in London, UK, who had the chance to review the findings of Wang Bo and Huang Diying, mentioned that both groups of scientists claimed that they were not aware of the other's investigations during the prepublication stage (Pennisi, 2018). Paul Selden, a paleontologist at the University of Kansas, USA, was intrigued by research insights on the origin and evolution of spiders in the Cretaceous Period that were published in the journal *Nature Ecology and Evolution,* as he said in scientific communication, "Interest stepped up about ten years ago when it was discovered this amber was mid-Cretaceous … Therefore, all the insects found in it were much older than first thought" (Wang et al., 2018). The detailed results from the two investigations on prehistoric arachnid specimens provided a clearer picture about the complicated anatomical formations of this prehistoric creature, like the spinnerets in this unique organism, helped evolutionary biologists claim it as the evolutionary predecessor of modern spiders. Jaison Bond, an evolutionary biologist at Auburn University in Alabama, USA, critically reviewed the publications of Wang Bo and Huang Diying and agreed with the argument for the etymological interpretation of this prehistoric "Chimeric spider" due to the presence of composite characteristic features such as the presence of the flagellum or long whippy taillike organ of a scorpion and typical spider characteristics such as a couple of fangs, four pairs of walking legs, spinnerets (the silk weaving appendage), pedipalps of male spiders (the sperm-injecting organ of a male), etc. in this unique creature led him infer that it was an ancestor of modern spiders (Wang et al., 2018; Pennisi, 2018). Elucidating the presence of the flagelliform taillike organ of the "Chimeric spider," Paul Selden said, "*Any sort of flagelliform appendage tends to be like an antenna ..... it's for sensing the environment. Animals that have a long whippy tail tend to have it for sensory purposes*" (Wang et al., 2018). In fact, scientists did not notice the taillike flagelliform organ in most modern-day spiders but discovered it in the relatives of modern-day spiders recognized as "Vinegaroons" (Wang et al., 2018). In their early investigations on the origin and evolution of spiders, Paul Selden and his associates came across a number of primitive, spiderlike arachnid fossils from the Devonian−Permian Period that proved the existence of spiderlike arachnids a few hundred million years earlier than the "Chimeric spider" existed on Earth (Wang et al., 2018). About his earlier works, Paul Selden has further said "*Once we recognized the difference between earlier arachnid that had a tail but not the spinnerets ... That's why the new one is really interesting, apart from the fact that it's much younger—it seems to be an intermediate form*" (Wang et al., 2018). From the age-dating technique, scientists determined the age of the fossilized amber as 100 million years. By referring to this data, Jason Bond has hypothesized further that the ancestors of modern spiders might have evolved much earlier (as the earlier investigations ascertained that prehistoric arachnids evolved around 420−305 mya) before the overlapping period of divergence of modern-day spiders evolved from the ancestral lineage, and the time of tentative period of evolution of "Chimeric spider" indicated that both evolutionary cousins might have coexisted for 200 million years before the extinction of "Chimeric spider," and possibly at that time, the divergence of the "modern-day spider" was set (Pennisi, 2018). Whether this newly discovered *C. yingi* had evolved with any silk-spinning and bug-trapping potential has remained an unresolved topic to evolutionary biologists, but scientists have hypothesized that most likely, the tropical rain forests in Myanmar (erstwhile Burma) and other parts of the world might have gone through a series of changes (e.g., climatic, environmental, atmospheric, physical) in the Cretaceous Period that altered the erstwhile predator—prey equilibrium, which supposedly put the domination of the scorpion members in the arachnid group of organisms on hold and favored the spiders to evolve with silk-spinning spiders and bug-trapping potential so they could roam around every ecosystem in the world in the post-Cretaceous regime. Whatever it is, it seems that the web-spinning potential might have played the cardinal role in the origin, evolution, and diversification of spiders in the course of time.

There is no doubt that spiders evolved around 400 mya as a terrestrial invertebrate predator, but scientists have found it difficult to determine (even after putting sincere effort into such investigations), as the phylogenetic relationships among spiders at inter- and intraspecific levels and the diversification dynamics of arachnids have revealed to form the spider tree of life. The diversity of the structural formation of webs jeopardizes the origin and evolution story of spiders, as it ranges from the "trampoline-like structure" spun by the sheet-web spider to the "filigree-like structure" spun by the orb-web spiders. The diverse array of spiders is broadly categorized into two groups, sticky orb-webs and nonsticky orb-webs. Erstwhile investigations on the DNA sequencing investigations on spiders and the interpretations of the findings in the 1980s—90s by evolutionary biologists have contemplated that both these distinct types of spiders evolved from a common stock, i.e., Orb-web spiders (Eck, 2018). Moving away from the adoption of conventional mode evolutionary investigations, Gustavo Hormiga, the eminent evolutionary biologist of George Washington University, USA along with a team of six experts in the evolutionary biology of arachnids, has tried to examine the validity of the "ancient orb-web hypothesis" that spiders evolved to spin the orb-web only once in their life. To determine whether spiders evolve to spin the orb-web only once or more than once, Gustavo Hormiga and his associates employed a composite approach to decipher spider evolution through phylogenomics, transcriptomics, and lineage diversification analysis, and they analyzed 2500 genes selected from 159 species of spider (Fernandez et al., 2018). The preliminary investigations of Hormiga and his colleagues, who successfully built spider family trees by sequencing the smaller sequence of DNA, observed that identical genetic markers seemed to be phylogenetically closely related to spiders with different genetic markers (Xu et al., 2015). To make an inference from the perspective of evolutionary biology, scientists used an experimental model based on one variable category, such as "orb-web to hunt" for the spider that does not build a web. Furthermore, team Hormiga drew a firm conclusion about intraspecific spiders, that the sticky orb-web is phylogenetically closely related, but spiders that make nonsticky orb-webs must have evolved from an ancestral lineage that never engaged in forming and using webs for hunting (Hormiga and Griswold, 2013). After their critical observations, Hormiga and his team noticed that there was no substantial evidence of the integration of the loss of foraging webs and the enhancement in the diversification rates, which led scientists to infer that factors like habitat heterogeneity or biotic interactions played the cardinal role in the diversification of modern spiders. Furthermore, scientists emphasized that key factors, such as the high extent of genomic differences in the main spider lineage, witnessed in that experiment played an important part in the upgradation of behavioral and sensory receptions as well as physiological and immunological upgradation and polyphonic determination (Fernandez et al., 2018). It was also determined by scientists that the extensive array of transcriptomics analysis supplemented with lineage diversification analysis had led scientists not only to draft a realistic backbone for the spider tree of life but also to nullify the "ancient orb-web hypothesis" of the 1990s as team Gustavo had come up with a new phylogenomic interpretation of origin and evolution of spiders, that clearly stated that "orb-webs evolved multiple times since the late Triassic—Jurassic" (Fernandez et al., 2018). The conventional interpretation of the transcriptomics data in the phylogenetic investigations of the spiders led arachnologists and evolutionary biologists to construe that the ancestral lineage of spiders did not have any potential to spin foraging web and it helped the scientists to contemplate that the spider webs evolved de novo (starting from the beginning) 10—14 times; whereas the orb web evolved 2 to 3 times (Coddington et al., 2019). But several molecular phylogeneticists have adopted an alternative method to analyze the transcriptomic data by adopting the strategies of character optimization algorithms, which yields "counterintuitive results," as it states that "foraging webs are primitive for spiders" and that webs have been lost ~5—7 times, and the orb web may have been lost 2—6 times (Coddington et al., 2019). So, the contemporary observations related to the advancement of spider phylogenomics and molecular genetics intrigued evolutionary biologists enough to contemplate a unique hypothesis of the evolution of spiders that in the course of evolution "spiders did not repeatedly gain but lose, foraging webs" (Coddington et al., 2019). However, many arachnologists and evolutionary biologists did not agree with Hormiga's interpretation on the origin and evolution of spiders, like Jason Bond, an evolutionary biologist at Auburn University, USA.

He firmly criticized the views of Hormiga (2017), as he commented that the way the interpretation had been presented, with the distinct structural formation and web architecture of the spiders against the backdrop of the molecular genetics and phylogenomics of spiders, made it appear that the orb-web forming and hunting traits of spiders had evolved around 10 times, which seemed to be an implausible hypothesis, and he said, "*I would say these authors [team Hormiga] have climbed pretty far out on to a limb … It will be interesting to see if it hold*" (Greenwood, 2018).

Although spiders are an ancient animal, and one of the most diverse belonging to the order Araneae, showing diversity in terms of genome size, karyotypes, and sex chromosomes. It intrigues molecular biologists to become involved in the investigations on its traits to observe and interpret its evolutionary journey over the last 400 million years, but real progress in this field of study is very limited. Arachnologists and evolutionary biologists have observed that the preliminary level of information of molecular phylogeny has primarily been gathered on the Entelegynae subgroup of the infraorder Araneomorphae (which is considered a diversified clade of Araneae) rather than accumulating any molecular data on the other major clade of spiders belonging to mesothelines, mygalomorphaes, haplogynae araneomorphaes, etc. (Král et al., 2019). Since the 1990s, a series of investigations into the domains of cytogenetics, molecular genetics, and genomics on spiders were conducted by several investigators, such as Maddison (1982) on the sex chromosomes of the male spider of *Pellens* sp.; the sex-linked fusion heterozygosity in the Australian huntsman spider, *Delena cancerides*, by Rowell (1985); the evolution of multiple sex chromosomes in the spider genus *Malthonica* by Král (2007); Sex chromosomes and molecular mechanism of meiosis in spiders by Araujo et al. (2012); karyotypes, sex chromosomes and meiotic division of spiders by Kořínková and Král (2013); and the duplication of the ancient genome of the house spider by Schwager et al. (2017), highlight many complex issues such as the diversity of karyotypes, evolution and diversity of spider genomes, genome duplication, and the formation and assemblage of multiple sex chromosomes (including multiple sex chromosomes) against the backdrop of the evolutionary journey of arachnids.

The preliminary findings on molecular footprints about the evolutionary journey of spiders intrigued Jiří Král, Department of Genetics and Microbiology, Charles University, Vinicna, Czech Republic to coordinate collaborative research on the spider genome evolution to decipher karyotype and genome changes related to the transition from standard (monocentric) to a holocentric (holokinetic) chromosome structure of the haplogynae araneomorphaes. Although the discovery of holokinetic chromosomes happened around 50 years ago, scientists have found that information on the existence and functions of these typical chromosomes in spider phylogeny are quite limited (Diaz and Saez, 1966). Now, it is an essential concern to delineate the importance of studying the holokinetic chromosomes in relation to evolutionary biology. From a cytogenetical perspective, the chromosomes do not have any center (i.e., centromere), so the centromere is not attached to any chromatid (Mola and Papeschi, 2006). Hence, kinetic activities might be shifted to any point, ranging from one end to the other ends of the chromosome. However, the accidental breaking of any holokinetic chromosome led to the movement of such breakaway part of a chromosome to the daughter cells to be part of the concerned genome (Král et al., 2019). So the evolutionary implications of the organisms having holokinetic chromosomes containing the higher potential of chromosomal variation in their karyotypic profile leads to the karyotypic evolution of the concerned organism or group of organisms. Although the possession of holokinetic chromosomes is not a common cytogenetical trait within a wide group of organisms, it was reported among 350,000 species and phylogenetic investigations further ascertained that holokinetic chromosomes evolved many protists, plants, and invertebrates more than several times (Bureš et al., 2013). Phylogenetic investigations have revealed that holokinetic and holocentric chromosomes can be found chiefly within four groups of the animal clade as follows (Král, 1994):

a. Acariform mites
b. Moths, butterflies, and caddis flies
c. Hemipteroid insects (includes some phylogenetically related orders) and
d. Some nematode lineages

In the phylogenetic investigation of arachnids concerning the availability of holokinetic chromosomes, evolutionary biologists deciphered that in the spiders, acariformes mites and buthid scorpions have evolved at least three times (Král et al., 2006; Schneider et al, 2017). Scientists have further observed that recombination, segregation, and chromosomal fusion frequency in the holokinetic chromosomes are very low (Mola and Papeschi, 2006). Cytogeneticists observed that the segregation of the holocentric chromosomes happened quite an unusual way as the microtubules of the spindle are generally attached to the surface of the chromosome toward the polar end, and the chromosomal segregation of the holocentric chromosomes happens uniquely, as an attachment of the microtubule at the chromosomal surface at the polar end in the first meiotic division resulted in a typical form of segregation of sister chromatid recognized as "inverted meiosis" (Marques and Pedrosa-Harand, 2016). Many scientists observed a different format of segregation of holocentric chromosomes, as the attachment of the microtubules happened to the chromosomal surface, particularly in the telomere region, and such attachment of microtubules at the telomeric end of the chromosome was assumed to be triggered by the kinetic activity, so that course of meiotic maneuvering during chromosomal segregation was scientifically recognized as "telokinetic meiosis" (Viera et al., 2009; Marques and Pedrosa-Harand, 2016). Based on the occurrence of holocentric chromosomes in a diversified and unrelated group of organisms, evolutionary biologists further hypothesized that spiders might have evolved from a very simple format of karyotypic evolution but the mode of its occurrence in the phylogenetically unrelated group of organisms. To explore and define such possibilities through the vertical trail of biological evolution, the contemporary group of evolutionary biologists tries to peep through the unusual kaleidoscope of evolution, where an opportunity of exploring the experimental opportunities through horizontal gene transfer protocol among the phylogenetically unrelated group of organisms has a limited scope (Král, 1994; Rogers, 2012).

The investigation on the occurrence of holokinetic chromosomes of the haplogyne araneomorphs revealed that around 5300 species belonged to 18 families, of which two families were like Dysderidae, most of the members of which have holokinetic chromosomes (Diaz and Saez, 1966). Biogeographical, paleontological, and phylogenetic investigations revealed that the genus Dysderoidea of the Dysderidae family, which had around 2700 species, and most species possessed holokinetic chromosomes, an ancestral lineage that presumably evolved between the Cretaceous (around 145 mya) and Triassic Periods (around 251.9 mya) (Izquierdo and Labarque, 2010; Penny and Selden, 2011). In the contemporary efforts of phylogenomic analyses of spider phylogeny, the family members of Dysderoidea are grouped into two phylogenetically close families such as Caponiidae (having 119 species) and Trogloraptoridae (having one species) nesting in the clade of proximity (World Spider Catalog http://wsc.nmbe.ch [15th August , 2020]). The phylogenomic investigations of the origin and evolution of holokinetic chromosomes in plants have been integrated into the considerable changes in genome size and changes in AT (purine—pyrimidine based pair adenine—thiamine) to GC (purine based pair guanidine—cytocine) in the coding region of the genome, and the higher GC content in the coding region of the human genome plays an important role in the evolution of modern human species on Earth (Bureš et al., 2013; Smarda et al., 2014). When contemporary evolutionary biologists and molecular genomicists investigated the genome size and GC content of the genome of the holokinetic group of haplogyne araneomorph, scientists observed that GC content in the genome of this specific group of arachnids was negligible or missing, which stalled the analysis and interpretation of the evolutionary role of holokinetic chromosomes of spiders to unfurl its history of origin and evolution. Undertaking a cladistics analysis helped draw a phylogenetic tree of the spider (Král et al., 2019). According to recent observations and interpretations of the phylogenetic investigations done by Král et al. (2019), the karyotype of Dysderoidea, the ancestral lineage of modern-day spiders, has three pairs of autosomes and a single X chromosome, and the karyotypic evolution of its holokinetic chromosomes leads to inverted meiosis of its sex chromosome and enhancement of 2n. Dr. Jiří Král and his fellow scientists (2019) noticed that Caponiids, the sister clade of Dysderoidea, have genomes with a high content of diploid chromosomes and sex chromosomes, which indicates the occurrence of polyploidy in the ancestral lineages of Caponiids. According to the observations of contemporary molecular phylogeneticists, holokinetic chromosomes would have formed as a result of fusions of multiple chromosomes and a substantial shrinkage in genome size (Král et al., 2019). At the end of contemporary phylogenomic investigations, Dr. Jiří Král and fellow scientists (Král et al., 2019) have come up with the proposition that the sex chromosomes might not stall a cardinal impediment to polyploidy that had been observed in the genomes among the higher group of organisms. Rather, the spiders had evolved with unique evolutionary karyotypic maneuvering that prompted overriding integration of copies of sex chromosomes or allosomes in the genome, and still enjoyed the karyotypic evolutionary advantages triggered by polyploidy-induced diversification of its gene to cope with each ecological niche to meet the survival challenges during its journey of evolution that most likely started about 400 million years ago.

References

Araujo, D., Schneider, M.C., Paula-Neto, E., Cella, D.M., 2012. Sex chromosomes and meiosis in spiders: A review, meiosis. Molecular mechanisms and cytogenetic Diversity, Andrew Swan. IntechOpen. https://doi.org/10.5772/31612. Available from: https://www.intechopen.com/chapters/30368.

Bureš, P., Zedek, F., Marková, M., 2013. Holocentric chromosomes. In: Leitch, I.J., Greilhuber, J., Dolezel, J., Wendel, J. (Eds.), Plant Genome Diversity. Physical Structure, Behaviour and Evolution of Plant Genomes, vol. 2. Springer, Heidelberg, pp. 187–208.

Coddington, J.A., Agnarsson, I., Hamilton, C.A., Bond, J.E., 2019. Spiders did not repeatedly gain, but repeatedly lost foraging webs. PeerJ 7, e6703. https://doi.org/10.7717/peerj.6703.

Diaz, M.O., Saez, F.A., 1966. Investigaciones citogenéticas sobre algunas especies de araneidos uruguayos. An. II Congr. Lat. Am. Zool. 2, 3–9.

Eck, A., April 30, 2018. Scientists Draw New Spider Web Family Tree. https://www.pbs.org/wgbh/nova/article/scientists-draw-new-spider-web-family-tree/.

Fernández, R., Kallal, R.J., Dimitrov, D., Ballesteros, J.A., Arnedo, M.A., Giribet, G., Hormiga, G., 2018. Phylogenomics, diversification dynamics, and comparative transcriptomics across the spider tree of life. Curr. Biol. 28 (9), 1489–1497.e5.

Garwood, R.J., Dunlop, J.A., Selden, P.A., Spencer, A.R.T., Atwood, R.C., Vo, N.T., Drakopoulos, M., 2016. Almost a spider: a 305-million- year-old fossil arachnid and spider origins. Proc. R. Soc. B. 283. https://doi.org/10.1098/rspb.2016.0125, 20160125.

Goulson, D., April 25, 2014. The Beguiling History of Bees [Excerpt]. https://www.scientificamerican.com/article/the-beguiling-history-of-bees-excerpt/.

Greenwood, V., April 26, 2018. A New Spider Family Tree Tries to Untangle the Evolution of Webs. http://nyti.ms/1MbHaRU.

Hormiga, G., 2017. The discovery of the orb-weaving spider genus Pinkfloydia (Araneae, Tetragnathidae) in eastern Australia with description of a new species from New South Wales and comments on the phylogeny of Nanometinae. Zootaxa 4311 (4), 480–490.

Hormiga, G., Griswold, C., 2013. Systematics, phylogeny, and evolution of orb-weaving spiders. Annu. Rev. Entomol. 59. https://doi.org/10.1146/annurev-ento-011613-162046.

Izquierdo, M., Labarque, F.M., 2010. Description of the female of Orsolobus pucara Forster & Platnick 1985, with comments on the functional morphology of the female genitalia in Dysderoidea (Araneae: Dysderoidea: Orsolobidae). J. Arachnol. 38 (3), 511–520. https://doi.org/10.1636/A09-071.1.

Kořínková, T., Král, J., 2013. Karyotypes, Sex Chromosomes, and Meiotic Division in Spiders. https://doi.org/10.1007/978-3-642-33989-9_12.

Král, J., 1994. Holokinetic (holocentric) chromosomes. Biol. Listy 59, 191–217.

Král, J., 2007. Evolution of multiple sex chromosomes in the spider genus Malthonica (Araneae: Agelenidae) indicates unique structure of the spider sex chromosome systems. Chromosome Res. 15, 863–879. https://doi.org/10.1007/s10577-007-1169-3.

Král, J., Musilová, J., Stahlavsky, F., Rezác, M., Akan, Z., Edwards, R., Coyle, F., Ribera, C., 2006. Evolution of the karyotype and sex chromosome systems in basal clades of araneomorph spiders (Araneae: araneomorphae). Chromosome Res. 14, 859–880.

Král, J., Forman, M., Kořínková, T., Lerma, A.C.R., Haddad, C.R., Musilová, J., Řezáč, M., Herrera, I.M.A., Thakur, S., Dippenaar-Schoeman, A.S., Marec, F., Horová, L., Bureš, P., 2019. Insights into the karyotype and genome evolution of haplogyne spiders indicate a polyploid origin of lineage with holokinetic chromosomes. Sci. Rep. 9, 3001. https://doi.org/10.1038/s41598-019-39034-3.

Maddison, W.P., 1982. New species of Habronattus and Pellenes jumping spiders (Araneae, Salticidae, Harmochirina). ZooKeys 646, 45–72. https://zookeys.pensoft.net/issue/1013/.

Marques, A., Pedrosa-Harand, A., 2016. Holocentromere identity: from the typical mitotic linear structure to the great plasticity of meiotic holocentromeres. Chromosoma 125. https://doi.org/10.1007/s00412-016-0612-7.

Mola, L., Papeschi, A.G., 2006. Holokinetic chromosomes at a glance. J. Basic Appl. Genet. 17, 17–33.

Pappas, S., June 20, 2013. Paleozoic Era: Facts & Information. https://livescience.com/37584-paleozoic-era.html.
Pappas, S., March 31, 2016. 305-Million-Year-Old 'Almost Spider' Unlocks Arachnid History. https://www.scientificamerican.com/article/305-million-year-old-almost-spider-unlocks-arachnid-history/.
Pennisi, E., February 5, 2018. Part Spider, Part Scorpion Creature Captured in Amber. https://www.sciencemag.org/news/2018/02/part-spider-part-scorpion-creature-cap tured-amber.
Penney, D., Selden, P.A., 2011. Fossil Spiders: The Evolutionary History of a Mega-Diverse Order. Monograph Series, vol. 1. Siri Scientific Press, Manchester, U.K, 128 pp.
Rogers, A.B., 2012. Gastric Helicobacter spp. in animal models: pathogenesis and modulation by extragastric coinfections. Methods Mol. Biol. 921, 175–188. https://doi.org/10.1007/978-1-62703-005-2_21.
Rowell, D.M., 1985. Complex sex-linked fusion heterozygosity in the Australian huntsman spider *Delena cancerides* (Araneae. Sparassidae).Chromosoma 93, 169–176.
Schneider, M.C., Mattos, V.F., Cella, D.M., 2017. The Scorpion Cytogenetic Database. Avaliable at. http://www.arthropodacytogenetics.bio.br/scorpionsdatabase/index.html.
Schwager, E.E., Sharma, P.P., Clarke, T., et al., 2017. The house spider genome reveals an ancient whole-genome duplication during arachnid evolution. BMC Biol. 15, 62. https://doi.org/10.1186/s12915-017-0399-x.
Selden, P., Corronca, J., Hünicken, M., 2005. The true identity of the supposed giant fossil spider Megarachne. Biol. Lett. 1, 44–48. https://doi.org/10.1098/rsbl.2004.0272.
Selden, P., Shear, W., Sutton, M., 2009. Fossil evidence of the origin of spider spinnerets and a proposed arachnid order. Proc. Natl. Acad. Sci. U. S. A. 105, 20781–20785. https://doi.org/10.1073/pnas.0809174106.
Šmarda, P., Bureš, P., Horová, L., Leitch, I.J., Mucina, L., E. Pacini, E., Tichý, L., Grulich, V., Rotreklová, O., 2014. Ecological and evolutionary significance of genomic GC content diversity in monocots. Proc. Natl. Acad. Sci. U. S. A. 111, E4096–E4102.
Switek, B., 2010. Megarachne, the giant spider that wasn't. Biology Letters. ScienceBlogs. 1 (1), 44–48. https://doi.org/10.1098/rsbl.2004.0272.
Viera, A., Page, J., Rufas, J., 2009. Inverted meiosis: the true bugs as a model to study. Genome Dyn. 5, 137–156. https://doi.org/10.1159/000166639.
Wang, B., Dunlop, J.A., Selden, P.A., Garwood, R.J., Shear, W.A., Muller, P., Lei, X., 2018. Cretaceous arachnid Chimerarachne yingi gen. et sp. nov. illuminates spider origins. Nat. Ecol. Evol. 2, 614–622. https://doi.org/10.1038/s41559-017-0449-3.
World Spider Catalog. http://wsc.nmbe.ch [15th August, 2020]. World Spider Catalog. Version 21.5. Natural History Museum, Bern online at https://doi.org/10.24436/2. Accessed on 27 Jan 2021.
Xu, X., Liu, F., Chen, J., Ono, H., Li, D., Kuntner, M., 2015. A genus-level taxonomic review of primitively segmented spiders (Mesothelae, Liphistiidae). ZooKeys 488, 121–151. https://doi.org/10.3897/zookeys.488.8726.

CHAPTER NINE

Evolutionary repertoire on silk spinning by spiders: Origin and Microevolution of spinnerets and role of the spidroin gene family

Abstract

Silk spinning is the key factor in understanding the ecology of spiders and their widespread dispersal and acclimation in different ecoregions of the entire terrestrial ecosystem on Earth in the course of expansive evolutionary diversification. Spider silk is primarily composed of modified proteins encoded by the multigene family. The phylogenetic diversity of spiders has not been analyzed by molecular phylogeneticists and genome biologists by examining the silk genes retained by its genome, till end of 20th century. Furthermore, the repertoire on "duplication and divergence of spidroin gene" and "gene tree—species tree reconciliation" has not yet fully convinced the interdisciplinary scientists (involved in decrypting the origin and evolution of spiders on Earth) due to a lack of evidence to correlate the key issues. These issues include understanding the origin and evolution of spinnerets in Araneae and the causes and effects of spidroin gene diversity to deliver a feasible explanation of the phylogenetic diversity of spiders on Earth.

Keywords: Evo-devo; Gene expression; Spider silk; Spidroin; Spinnerets.

Silk spinning is essential to spider ecology and has had a key role in the expansive diversification of spiders. Silk is composed primarily of proteins called spidroins, which are encoded by a multi-gene family.

James Starrett et al.[1]

Whether it's within the domain of science & engineering (such as mechanical engineering or architecture), arena of ancient history, or field of cultural anthropology, the spider's silk-spinning potential has mesmerized

[1] Jessica Garb, Amanda Kuelbs, Ugochi Azubuike, and Cheryl Hayashi.

Spider Evolution
ISBN: 978-0-323-90041-6
https://doi.org/10.1016/B978-0-323-90041-6.00006-1

humankind since the dawn of civilization. The production of silk fibers from the opisthosomal glands is the characteristic feature of spiders (Araneae). Scientists have observed that web-spinning spiders use special appendages called spinnerets (found on the posterior part of their bodies attached to the opisthosoma or abdominal segment) for weaving the cowebs spun with spider-silk threads (Selden et al., 2008). The emergence of silk fiber takes place through spigots (modified setae) integrated into the spinnerets (modified appendages) near the internal silk glands of spiders (Selden et al., 2008). These unique appendages adhere to the spider-silk producing glands to extrude and maneuver silk in liquid form from the silk glands, dry it in the air to form silk threads, and use it to spin webs. The earliest record of spigots is from the prehistoric fossiliferous representative of 386 million years ago (mya) belonging to Araneae, *Attercopus fimbriungus* from the Middle Devonian strata of Gilboa, New York (Shear et al., 1989). Paleontologists have meticulously studied the slide (slide no. 334.1b.AR34) of this earliest fossilized specimen of a fusiform spinneret of *A. fimbriungus* that bore around 20 spigots arrayed at the medial surface of the spinneret (Shear et al., 1989; Selden et al., 1991). Furthermore, scientists have opined that the fossilized spinnerets have close morphological and anatomical resemblance with the posterior median spinnerets of the primitive group of spiders belonging to Mesothelae (Selden et al., 2008). Although there is a great deal of importance in understanding the mystery of spiders' web-building skills linked to the functional aspects of their spinnerets, the fact is that there is a lack of substantial progress in research related to the origin and evolution of spinnerets in the phylum of arachnids.

Arachnologists have observed that commendable findings have been made in the field about the morphological and anatomical aspects of spider spinnerets. These findings include knowledge on the chemical and mechanical properties and identities of spider silks, their widespread applications in various applied fields of science and engineering, ranging from chemical engineering to fiber technology to medical technology to contemporary nanotechnology. In the contemporary frontier research on arachnology, interdisciplinary research has been initiated by developmental biologists, molecular geneticists, and genome biologists. Those scientists have strived to track down molecular footprints on the origin and evolution of spinnerets and their unique phase in the embryonic development of spiders. To move research forward quickly, scientists reviewed a barge of hypotheses on the origin and diversification of spider spinnerets to construct an evo-devo pathway to make a good start. However, before embarking on the trail of the origin and evolution of spinners of arachnids, we need to grasp basic ideas about the anatomy of spiders, the well-diversified clade of around 48,000 species that has occupied nearly all the terrestrial niches in the world (World Spider Catalog, 2019). Arachnologists had contemplated that the prehistoric predecessor of extant spiders used the thin films of "flimsy silk filaments" to build shelters or "diving bells to survive flooding" during the prehistoric time on Earth when most of its ecosystems had gone through a major paradigm shift coming across the regime of transmigration of the aquatic biosphere to the terrestrial biosphere (Vollrath, 2010). Evolutionary biologists contemplated further that an arms race "between insect prey and arachnid predator" has promoted the emergence of complex behavioral traits of spiders and expedited its higher potential to produce a diversified range of silk fibers (Vollrath, 2010). Regarding shedding light on the origin and evolution of spider silk by arachnids, Michael R. Gray, the eminent arachnologist, expressed his opinion optimistically in 1978, when he said, "… *the evolution of spider silk has been an event comparable in importance to the evolution of flight in the insects, or warm-bloodedness in the vertebrates*" (Foelix, 2010).

Spiders are well recognized as predominantly diversified, nonvertebrate, arthropod clade belonging to the distinct taxonomic group of animals, called Arachnida (Araneae), are devoid of any endoskeleton or bones but protected in the outer environment by an exterior hard shell exoskeleton made up of chitin. Although a great number of morphoanatomical and physiobehavioral diversities have been witnessed among arthropods, spiders are typically characterized by features such as the presence of chelicerae adhered to the venom glands and the presence of modified tarsi in the pedipalps that male spiders use to transfer sperm to their female counterparts during mating (Wheeler et al., 2017). Most insects have three body segments—head or cephalic part, chest or cephalic part, and abdomen or stomach part—whereas spiders are characterized by a body with two distinct parts:

a. Cephalothorax: The front part of the body comprising the fused part of the head, or cephalic part; and the chest, or thorax part; fused with an indistinct cervical groove and a notch of depression to which the abdominal part is attached by several muscles. The important organs of arachnids such as venom glands, the central nervous system in the brain, the sucking stomach, the intestine, etc., are found in the cephalothorax. The key appendages of the cephalothorax are as follows:

i. Four pairs of segmented legs (having seven segments) comprising the femur (thigh), patella (knee), tibia(lower leg), metatarsus (foot), and tarsus (toes) with tarsal toenails
ii. A pair of pedipalps (functionally recognized as arms) resembling legs but devoid of the metatarsus
iii. A pair of chelicerae with hinged fangs

b. Abdomen (opisthosoma): The distal part of the body attached to the cephalothorax has been recognized as the abdomen, which houses some appendages and key functional organs such as the heart, ovaries, respiratory organs, and guts, and the primary functional appendages of the abdominal part of spiders are as follows:
i. Book lungs or archaic respiratory organ in arachnids
ii. Four to six spinnerets or silk-producing appendages yielding a variety of silk threads
iii. The anal tubercle, noticed conspicuously above the spinnerets, through which excreta are discarded from time to time

Although some distinctive features characterize spiders, this distinct group of arthropods is recognized as arachnids based on whether its members have the potential to yield silk threads and the manipulating abilities to weave it. The meticulous ecological and experimental observations of evolutionary biologists have led them to hypothesize that the silk spinning and manipulating abilities of this clade Araneae by using the spinning appendages, called "spinnerets," adhered to the silk glands should be considered as a synapomorphic instance of arachnids that existed as archaic predecessors of the present-day spiders, the features, originated and evolved in its root-stock millions of years ago (Wang et al., 2018). Biochemically, silk threads are made up of fibroin, an insoluble proteinaceous sap, contained by the various morphologically and physiologically distinguished silk glands inside the opisthosoma and spinnerets; the silk-producing appendages can be found in the fourth and fifth abdominal segments (Mariano-Martins et al., 2020). To define the diversified nature of silk glands in spiders, arachnologists have tendered an example of *Trichonephila clavipes*, an orb-weaving spider that possesses seven types of morphologically and physiologically differentiated silk glands: minor ampullate, major ampullate, aciniform, tubuliform, aggregate, piriform, and flagelliform. The distinct nature of gene expression in these diversified silk glands, leading to the production of distinct spidroins—proteins yielding specific silk threads exuded from distinct silk glands (Rising and Johansson, 2015; Mariano-Martins et al., 2020). The specific kind of spinnerets are adhered to the specific type of silk glands in one end and opened up in the microscopic silk extruding from the external appendage called spigots through which the silk threads come out to the environment. Arachnologists have observed that the position of the spinnerets in a specific segment of the opisthosoma (either its O3, O4, or O5 segment) would help them classify its phylogenetic categories (Pechmann et al., 2010; World Spider Catalog, 2019). The spinnerets and spigots of the concerned spiders worked together or independently to put together distinct silk threads to coordinate multifarious tasks ranging from making shelters, communicating by transmitting vibrations or emitting pheromones (for mating), engaging in predation (foods), and dispersing further by forming silk balloons and lines (Mortimer et al., 2019).

The origin and evolution of spinnerets in spiders have gone through several bottlenecks of the evolutionary labyrinth of hypotheses. Around the 19th century, some evolutionists hypothesized that spiders had gone through the evolutionary cascade, wherein their prosoma legs ceased functioning as locomotory organs and transformed into spinnerets (Hilbrant, 2008). It is quite interesting to witness the observations in the developmental study by Pechmann and Prpic (2009), as they observed that the spinneret in the posterior segment of *A. geniculata* had identical leglike gene expression and further indicated the homology[2] between spinnerets and legs. Arachnologists have contemplated that the origin and evolution of spinnerets in spiders paved the way for the origin and evolution of silk production, and a group of evolutionary biologists has supported this hypothetical view. Those scientists referred to spigotlike appendages in the fossilized predecessors of the extant spiders, but these paleontological representatives are devoid of spinnerets, and it has failed to fit in with earlier observations (Shulz, 1987; Selden et al., 2008). A group of evolutionary biologists led by Huang et al. (2018) referred to the spinneret-bearing fossils of *Chimerarachne yingi*, a prehistoric palaeontological representative of protospider (apparently looking like a transitional animal structured between a scorpion and spider), to claim that the origin and evolution of spinnerets did not necessarily originate and likely evolve in the archaic spiders belonging to Araneae. This has raised a valid question among evolutionary biologists of whether the protospider should be considered an archaic progenitor of

[2] Homology: In terms of molecular biology, a homologous protein and gene (A DNA sequence) or homolog has been referred to as a gene (encoding the protein) inherited by two species having a common ancestor. Molecular biologists have contemplated that the shared ancestry of the two segments of DNA happened due to either speciation (orthologs) or gene duplication (paralogs).

the spider and whether it should be categorized within Araneae. The existence of spinnerets in the extant representative of spiders would lead evolutionary biologists to conclude that the presence of spinnerets in the fossiliferous representative of spiders helped them hypothesize that the origin and evolution of spinnerets must be traced back to the last common ancestor of Araneae, which supposedly appeared on Earth around 305 mya during the Carboniferous Period (Selden et al., 2014; Wang et al., 2018). However, several primary hypothetical explanations on the origin and evolution of spinnerets are as follows:

1. Shulz (1987), Hilbrant (2008), Pechmann and Prpic (2009) have propounded there is a homology between the spinnerets and the legs of the spider, mainly the posterior spinnerets of *Acanthoscurria geniculata.* Scientists have meticulously studied the gene expression of like organs of spiders and the developmental biology of the embryonic stage of spiders and contemplated further that the spinnerets emerged from the embryonic limb buds.
2. Averof and Cohen (1997) have tendered a shred of evidence on the homology of wings of insects and crustacean gills based on their observation, as they have witnessed an identical gene expression of pdm/ nubbin genes and apterous genes responsible for the expression of crustacean epipods (gills) and insect wings. In the case of the developmental stage of the embryo of the spider *Cupiennius salei*, scientists have noticed that both genes are responsible for gene expression of book lungs, tubular trachea, and spinnerets primordia (Damen et al., 2002). Moreover, Damen et al. (2002) have observed that the gene expression of chelicerates in spiders is identical to the other groups of arthropods within which it nested.
3. Damen et al. (2002), Selden et al. (2008) have observed the evidence of similar gene expressions in the book gills of *Limulus polyphemus* (horseshoe crab). Based on fossil evidence, Selden et al. (2008) have propounded that the origin and evolution of spinnerets in spiders (e.g., *A. fimbriungus*) have been triggered by the "gene expression reactivation in ventral plates of a uraraneida-like spider ancestor," and expression of the Distal-less (Dll) gene has supposedly regulated such gene expression (Mariano-Martins et al., 2020).
4. Clarke et al. (2015) have contemplated the hypothetical clarification of origin and evolution of spinnerets in spiders that supposedly happened as a result of the expression of the genes contained in its silk glands, which is likely to have happened as a result of "spider-specific gene

duplication." It has further been extrapolated that an event of retention of paralog[3] supposedly took place after the gene duplication events (Mariano-Martins et al., 2020).

Though the functional role-play of the genes expressed during the "spinneret embryonic development" has proved enigmatic to molecular geneticists, molecular biologists and evolutionary geneticists have been able to trace back the molecular pathways of gene expressions to demonstrate the origins and evolution of spinnerets in arachnids (Selden et al., 2008; Clarke et al., 2015). To demonstrate the gene expression related to the origin and evolution of spinnerets in arachnids, scientists have experimented with in situ hybridization to present expression of the genes: Dll, dachshund, extradenticle, and homothorax are noticed along with the expression of genes like pdm/nubbin and apterous genes in the expression of primordial appendages, which is an integrated issue to understand the ontogeny of spiders (Prpic et al., 2003; Pechmann and Prpic, 2009). A unique observation by evolutionary and developmental biologists is that the spinneret limb buds in the embryonic developmental stage of spiders mature as spinnerets in adult spiders, regulated by expression of the gene Dll (Mariano-Martins et al., 2020). Furthermore, developmental biologists have noticed that in the embryonic developmental phase of *A. geniculata*, pairs of spinneret primordial appendages have been observed in the O4 and O5 segments of the opisthosoma, but an expression of the Dll gene in the O5 segment regulates the growth of the spinnerets primordial appendages into functional spinnerets in the adult spider, responsible for web-spinning process (Pechmann and Prpic, 2009).

Contemporary morphological, embryonic developmental, and molecular genetic investigations on the arachnid species *C. salei* have recently been made by Hilbrant (2008), and he has meticulously studied the nature of gene expressions of the homologous genes of *C. salei*: AP-2 (activator protein 2), omb, H15, pairberry, and wingless. It is an amazing result that expression has been witnessed by the transcription factor AP-2 among the five homologs. An expression of AP-2 was initially witnessed in the spinneret limb buds in O4 and O5 opisthosoma but gradually diminished from O4. Hilbrant (2008)

[3] Paralog: During the occurrence of gene duplication in the same genome, several gene copies evolved, known as paralogs, whereas orthologs evolved as a result of speciation. When orthologous genes keep the gene functions unchanged in reference to the archaic gene profile, the paralogous genes have been emerged with development of different functions due to absence of selective pressure on one copy of the duplicated gene.

has further shared his observation to track down the molecular footprints of the evo-devo process of spinnerets in arachnids, as he has contemplated that the slow-growing spinneret limb-bud in the O4 segment of the abdomen has given rise to short-sized anterior spinnerets, whereas the limb bud in the O5 segment has differentially grown further to form long—sized median and posterior spinnerets.

Very recently, molecular geneticists have observed that most of the genes in the Hox clusters in the *Parasteatoda tepidariorum* genome are considered duplicated, which has led scientists to contemplate that similar pieces of evidence of duplication of the Hox cluster (that lead emergence of paralogs of each duplicated Hox gene) supposedly happened in another group of spiders, and these events of gene duplication paved the way for evolutionary divergence of scorpions and spiders (Schwager et al., 2017). Earlier molecular geneticists such as Abzhanov et al. (1999); Hughes and Kaufman (2002), who have engaged with an expression of Hox genes of spiders—abdominal-A and Ultrabithorax—noticed that these genes played a critical role in the emergence of spinnerets from the abdominal segment. At the same time, scientists have observed that the two Hox genes functionally repress expression of the Dll gene within the abdomens of insects, resulting in the prevention of appendages like spinnerets in the abdominal part of insects (Mariano-Martins et al., 2020).

Arachnologists have considered that silk weaving is the main criterion in studying the spider ecology (as the taxonomically diverse spider species are ecologically diverse in terms of origin and the evolution of arachnid silks) that played the key role in the widespread diversification of spiders in terrestrial ecosystems throughout this blue planet. Arachnids are efficient in constructing various lightweight spider-nets, which enable spiders to inhabit diverse ecological niches across the wide range of terrestrial habitats on Earth (Brunetta and Craig, 2010). It has generally been observed that an individual spider is capable of spinning a wide array of silk fibers, and each task-specific silk fiber (e.g., egg-case construction, prey wrapping, or web assembly) has emerged from the distinctive type of its abdominal silk glands (Hinman and Lewis, 1992). Biochemically, silk is made up of one or more modified proteins recognized as spidroins (a contraction of spider fibroins) that are genetically encoded by "multigene paralogs," gene duplication and divergence events, or a combination of all factors (Gatesy et al., 2001; Garb et al., 2007). They are mainly found in the orb-weavers (Orbiculariae) and belong to the group of "true spiders" or suborder Araneomorphae (Hinman and Lewis, 1992; Guerette et al., 1996; Gatesy et al., 2001). In their unique literature, titled "Spider Silk: Evolution and 400 million years of Spinning, Waiting, Snagging and Mating," Brunetta and Craig (2010) have eloquently narrated the complex nature of proteins that make up the spider silk fibers that have been studied by molecular biologists and biochemists in the dry state for years and gave preliminary ideas on protein foldings, and it helped in understanding the diversification of silk threads that have gone through selection over millennia.

To understand spider silk diversification, analyzing the evolutionary relationships of spidroins is essential. Spidroins exhibit a polymeric organization of their repetitive sequence, so the evolutionary relationship of spidroins has primarily been determined from their nonrepetitive carboxy (C)- terminal domains (having limited character data). A few spidroin amino (N)-terminal domains are difficult to obtain, but these amino-terminal domains have supposedly contained extremely important information for phylogenetic reconstruction of diversification of spiders (Garb et al., 2010). Molecular phylogeneticists have used phylogenetic information for reconstruction and diversification of spider silks, and scientists have used the silk-gland expression data or "expressed sequence tags" (ESTs) from divergent species to assess the functional significance and to understand the phylogenetic utility of spidroin amino-terminal domains (Garb et al., 2010). In their meticulous phylogenetic investigations, Garb et al. (2010) have reported 11 extra spidroin N-termini sequenced by 1900 silk-gland cDNAs prepared from nine species, and these nine species have presumably emerged from a prehistoric ancestor evolved around 240 mya (Garb et al., 2010). Molecular phylogeneticists have further noticed that with respect to the hypervariable repetitive regions of the carboxy (C) terminal domain of spidroin, its amino (N) terminal domains have retained an identical resemblance in "sequence identity," "tentative secondary structure," the extent of "hydrophobicity," etc. Molecular geneticists have coordinated separate and composite phylogenetic analyses of C-terminal domains and their corresponding N-terminal domains, and scientists have noticed that the composite phylogenetic analysis yields the ideal reconstruction of the phylogenetic tree (Garb et al., 2010). Furthermore, scientists have argued that the N-termini of spidroin can provide more support and less ambiguity than its C-termini flank. Based on the findings of their phylogenetic investigation, Garb et al. (2010) have concluded that a nonrepetitive N-terminal domain, considered to be the universal attribution of spidroin proteins, has supposedly been retained from the origin of spider silk on Earth. However, the phylogenetic analyses of the "conserved N-and C-terminal domains" of spidroins emerged with surged radiation in the spidroin gene family entailed by the random occurrence of gene duplications along with the high extent of diversification of repetitive structural sequences, which leads to structural, functional, and mechanical diversities in spider silk fibers (Garb et al., 2010).

Contemporary research on the origin and evolution of a diverse array of silks has broadly been contemplated as a result of duplication and divergence of spidroin genes in intensive research investigations made by arachnid genomicists (Starrett et al., 2012). The examination of molecular footprints of the spidroin gene family extracted from members of orbicularian spiders demonstrated that spidroin proteins are fairly long (around 15 kb) and comprise highly repetitive DNA sequences encoding this protein (Starrett et al., 2012). Though various amino acids (responsible for the formation of diverse silk threads) have been noticed by scientists in the protein-encoding region in the spidroin gene family, the preponderance of three amino acids—alanine, glycine, and serine—has been noticed by molecular biologists (Starrett et al., 2012). In recent molecular phylogenetic research on egg case proteins (ECPs), cysteine-enriched "second-class silk proteins" have been solely extracted from the egg cases of the western black widow (*Latrodectus hesperus*), so the evolutionary footprint of this typical tubuliform spidroin has been enigmatic until now (Hu et al., 2005, 2006). Scientists have observed that in terms of diversification of the silk weaving potential of spiders, orbicularian members are phylogenetically more advanced than spiders belonging to mesothele (spiders with segmented bodies), mygalomorh (tarantulas, trapdoor spiders, etc.), and paleocribellate spiders, as orb spiders are can yield as many as seven types of task-specific silk fibers (from morphologically distinct glands), whereas the members of mesothelin and mygalomorphs could produce very basic type substrate-borne silk fibers from their morphologically indistinct glands (Hu et al., 2006). Molecular phylogeneticists have further observed that the divergence of silk genes supposedly took place around 380 mya from different silk-weaving spiders but is absent in the members of Mesothelae (Starrett et al., 2012). From such a comparative phylogenetic profile, molecular genomicists have determined that the diversity of the silk genes in Mesotheles and Mygalomorphs is less likely in orbicularians. In recent phylogenomic investigations on the evolution and diversity of silk genes in spiders, James Starrett, along with four research colleagues, was able to synthesize cDNA by extracting DNA sequences from some mygalomorphs, a mesothele, and a nonorbicularian aranemorph, and the results of their investigations simply amazed them as they traced out ECP homologs in the representatives of Mesothele. Hence, the research findings of Starrett et al. (2012) have encouraged them to hypothesize that the molecular footprint of ECP is present in the genome of the common ancestral lineage of the extant spiders and might not have originally emerged as tubuliform spidroins (Starrett et al., 2012). Several recent phylogenetic investigations such as the reconstruction of the spidroin gene tree and reconciliation of gene tree/species tree analyses have further indicated that diversification of spidroins likely took place before the split of Mygalomorph and Araneomorph, and Mygalomorphs have emerged with the potential of yielding several spidroins, and the event of duplication of silk genes in arachnids took place after the phylogenetic split between Mesothelae and Opisthothelae (Garb et al., 2007; Starrett et al., 2012). Contemporary research on the origin and evolution of spinnerets and silk spiders has gone through a labyrinth of hypotheses. This research has supposedly been precisely supplemented by genetic and developmental data to crack the mystery supposedly initiated around 400 mya, and can deliver a feasible answer, yet research observations are sparsely scattered, and the knowledge gaps have remained relatively pertinent to this point.

References

Abzhanov, A., Popadic, A., Kaufman, T.C., 1999. Chelicerate Hox genes and the homology of arthropod segments. Evol. Dev. 1, 77–89. https://doi.org/10.1046/j.1525-142x.1999.99014.x.

Averof, M., Cohen, S.M., 1997. Evolutionary origin of insect wings from ancestral gills. Nature 385, 627. https://doi.org/10.1038/385627a0.

Brunetta, L., Craig, C.L., 2010. Spider Silk: Evolution and 400 Million Years of Spinning, Waiting, Snagging, and Mating. Yale University Press, USA, 248 pp.

Clarke, T.H., Garb, J.E., Hayashi, C.Y., Arensburger, P., Ayoub, N.A., 2015. Spider transcriptomes identify ancient large-scale gene duplication event potentially important in silk gland evolution. Genome Biol. Evol. 7, 1856–1870. https://doi.org/10.1093/gbe/evv110.

Damen, W.G., Saridaki, T., Averof, M., 2002. Diverse adaptations of an ancestral gill: a common evolutionary origin for wings, breathing organs, and spinnerets. Curr. Biol. 12, 1711–1716. https://doi.org/10.1016/S0960-9822(02)01126-0.

Foelix, R.F., 2010. Biology of Spiders, third ed. Oxford University Press, Oxford, U.K.

Garb, J.E., DiMauro, T., Lewis, R.V., Hayashi, C.Y., 2007. Expansion and intragenic homogenization of spider silk genes since the Triassic: evidence from Mygalomorphae (tarantulas and their kin) spidroins. Mol. Biol. Evol. 24 (11), 2454–2464.

Garb, J.E., Ayoub, N.A., Hayashi, C.Y., 2010. Untangling spider silk evolution with spidroin terminal domains. BMC Evol. Biol. 10, 243. https://doi.org/10.1186/1471-2148-10-243.

Gatesy, J., Hayashi, C.Y., Motriuk, D., Woods, J., Lewis, R.V., 2001. Extreme diversity, conservation, and convergence of spider silk fibroin sequences. Science 291, 2603–2605.

Guerette, P.A., Ginzinger, D.G., Weber, B.H., Gosline, J.M., 1996. Silk properties determined by gland-specific expression of a spider fibroin gene family. Science 272, 112–115.

Hilbrant, M., 2008. Development and Evolution of the Spider Silk Production System. Dissertation. University of Cologne, Köln.
Hinman, M.B., Lewis, R.V., 1992. Isolation of a clone encoding a second dragline silk fibroin. J. Biol. Chem. 267 (27), 19320–19324.
Hu, X., Kohler, K., Falick, A.M., Moore, A.M.F., Jones, P.R., et al., 2005. Egg case protein-1. J. Biol. Chem. 280 (22), 21220–21230.
Hu, X., Kohler, K., Falick, A.M., Moore, A.M.F., Jones, P.R., et al., 2006. Spider egg case core fibers: trimeric complexes assembled from TuSp1, ECP- 1, and ECP-2. Biochemistry 45, 3506–3516.
Huang, D., Hormiga, G., Cai, C., Su, Y., Yin, Z., Xia, F., et al., 2018. Origin of spiders and their spinning organs illuminated by mid-Cretaceous amber fossils. Nat. Ecol. Evol. 2, 623. https://doi.org/10.1038/s41559-018-0475-9.
Hughes, C.L., Kaufman, T.C., 2002. Hox genes and the evolution of the arthropod body plan1. Evol. Dev. 4, 459–499. https://doi.org/10.1046/j.1525-142X.2002.02034.x.
Mariano-Martins, P., Lo-Man-Hung, N., Torres, T.T., 2020. Evolution of Spiders and Silk Spinning: mini review of the morphology, evolution,and development of spiders' spinnerets. Front. Ecol. Evol. 8, 109.
Mortimer, B., Soler, A., Wilkins, L., Vollrath, F., 2019. Decoding the locational information in the orb web vibrations of Araneus diadematus and Zygiella x-notata. J. R. Soc. Interface 16, 20190201. https://doi.org/10.1098/rsif.2019.0201.
Pechmann, M., Khadjeh, S., Sprenger, F., Prpic, N.M., 2010. Patterning mechanisms and morphological diversity of spider appendages and their importance for spider evolution. Arthropod Struct. Dev. 39, 453–467. https://doi.org/10.1016/j.asd.2010.07.007.
Pechmann, M., Prpic, N.M., 2009. Appendage patterning in the South American bird spider Acanthoscurria geniculata (Araneae: Mygalomorphae). Dev. Genes Evol. 219, 189–198. https://doi.org/10.1007/s00427-009-0279-7.
Prpic, N.M., Janssen, R., Wigand, B., Klingler, M., Damen, W.G., 2003. Gene expression in spider appendages reveals reversal of exd/hth spatial specificity, altered leg gap gene dynamics, and suggests divergent distal morphogen signaling. Dev. Biol. 264, 119–140. https://doi.org/10.1016/j.ydbio.2003.08.002.
Rising, A., Johansson, J., 2015. Toward spinning artificial spider silk. Nat. Chem. Biol. 11, 309. https://doi.org/10.1038/nchembio.1789.
Schwager, E.E., Sharma, P.P., Clarke, T., Leite, D.J., Wierschin, T., Pechmann, M., et al., 2017. The house spider genome reveals an ancient whole- genome duplication during arachnid evolution. BMC Biol. 15, 62. https://doi.org/10.1186/s12915-017-0399-x.
Shulz, J.W., 1987. The origin of the spinning apparatus in spiders. Biol. Rev. 62, 89–113.
Selden, P.A., Shear, W.A., Bonamo, P.M., 1991. A spider and other arachnids from the Devonian of New York, and reinterpretations of Devonian Araneae. Palaeontology 34, 241–281.
Selden, P.A., Shear, W.A., Sutton, M.D., 2008. Fossil evidence for the origin of spider spinnerets, and a proposed arachnid order. Proc. Natl. Acad. Sci. U. S. A. 105 (52), 20781–20785. https://doi.org/10.1073/pnas.0809174106.
Selden, P.A., Shcherbakov, D.E., Dunlop, J.A., Eskov, K.Y., 2014. Arachnids from the carboniferous of Russia and Ukraine, and the permian of Kazakhstan. Paläontol. Z. 88, 297–307. https://doi.org/10.1007/s12542-013-0198-9.
Shear, W.A., Palmer, J.M., Coddington, J.A., Bonamo, P.M., 1989. A Devonian spinneret: early evidence of spiders and silk use. Science 246, 479–481.
Starrett, J., Garb, J.E., Kuelbs, A., Azubuike, U.O., Hayashi, C.Y., 2012. Early events in the evolution of spider silk genes. PLoS One 7 (6), e38084. https://doi.org/10.1371/journal.pone.0038084.
Vollarth, F., 2010. Silk evolution untangled. Nature 466, 319.

Wang, B., Dunlop, J.A., Selden, P.A., Garwood, R.J., Shear, W.A., Müller, P., et al., 2018. Cretaceous arachnid Chimerarachne yingi gen. et sp. nov. illuminates spider origins. Nat. Ecol. Evol. 2, 614. https://doi.org/10.1038/s41559-017-0449-3.
Wheeler, W.C., Coddington, J.A., Crowley, L.M., Dimitrov, D., Goloboff, P.A., Griswold, C.E., et al., 2017. The spider tree of life: phylogeny of Araneae based on target-gene analyses from an extensive taxon sampling. Cladistics 33, 574–616.
World Spider Catalog, 2019. World Spider Catalog. Natural History Museum, Bern version 20.5.

CHAPTER TEN

Adaptation and Evolutionary changes of spiders on planet Earth against the backdrop of geological time

Abstract

The biobehavioral evolution of spiders is inherently linked to its characteristic silk fiber production and the web-spinning architecture that results. Spiders have thus emerged as a diverse taxon occupying a niche as a key predator of insects as corradiated prey on Earth. The empirical analysis of the paleontological data on spiders and insects gradually evolved through geological regimes and revealed the exponential growth of both of these under the influence of adaptive radiation. Based on the comparable rates of diversification in both, evolutionary biologists have claimed that both are corradiated. As the radiation of insects and spiders happened around 100 million years ago before the radiation of angiosperms, it is perceived that insect evolution was lesser influenced by flowering plants, than triggered by spider predation. The diversity of spiderwebs played a cardinal role in driving this coevolutionary arms race.

Keywords: Coevolution; Insects; Silk; Spiders; Spiderweb.

> *Spiders' silks and webs have made it possible for this diverse taxon to occupy a unique niche as the main predator for another, even more, diverse taxon, the insects. Indeed, it might be that the spiders, which are older, were a major force driving the insects into their diversity in a coevolutionary arms race. The spiders' weapons were their silks and here we explore the evidence for the evolution of silk production and web building as traits in spider phylogeny.*
>
> ***Fritz Vollarth and Paul Selden***

Spiders have emerged with a wide range of morphological diversity, ranging from an enormous body like that of hairy mygalomorphs to the tiny pinhead size of oonopids; the number of eyes varied from two to eight from species to species; the main locomotory organs, the size of the hydraulic legs varied from very long to very short; the respiratory organs varied from the trachea to lungs among these fearsome creatures (Foelix, 1996). Arachnologists have observed that spiders' ecological and behavioral traits

Spider Evolution
ISBN: 978-0-323-90041-6
https://doi.org/10.1016/B978-0-323-90041-6.00001-2

have broadly been influenced by the presence of morphological, anatomical, and neurological characteristics (Snodgrass, 1952; Bristowe, 1958; Barth, 1985; Nentwig, 1986; Vollarth, 1992). From the perspective of adaptation, spider silk is a unique representation of behavioral traits in spiders, whether spinning a simple format of weblike "trailing a safety line" or forming a complex type of orb-web that indicates the advanced lifestyle of the concerned species. Although the structural diversity of spiderwebs is engineered by diverse kinds of silk thread, produced from distinct silk glands in different spider species, yet the silk type and development of specific silk threads have broadly been able to indicate the lifestyle of that spider. Hence, arachnologists have tried to put together some identifying features of spiders, such as their morphological, anatomical, biobehavioral, and neurological traits along with the applied features, such as web engineering (two dimensional and three-dimensional geometrics), silk thread use, and manipulation patterns, so as to achieve an advanced level of taxonomic analysis that helped them define spider phylogeny based on cladistics analysis (Witt et al., 1968; Eberhard, 1981; Shear, 1986; Coddington and Levi, 1991; Krink and Vollarth, 1997; Opell and Bond, 2001). The meticulous study in morphology and anatomy (by studying the intricate body shapes of the exterior and structural details of its interior) along with the physiological traits of spigots, silk-spinning glands, actions of sensory organs, the specific role of claws, etc. have helped arachnologists understand its behavioral traits (use of diverse nature of silk threads in weaving the specific types of the web), which further elucidated the phylogenetic background of the spider concerned. The availability of phylogenetic context combined with the contemporary analysis on its ecological data have helped phylogeneticists trace the evolutionary history of the ancestors of the extant group of spiders (Eberhard, 1981; Barth, 1985; Vollarth, 1987; Opell, 1989; Townley et al., 1993). For instance, when the availability of the paleontological history of spiders is incomplete, observation of the web architecture by the spider is an important steppingstone to track down its evolutionary history. Although it is broadly predictive, arachnologists have made an effort to infer the behavioral traits by analyzing morphological and anatomical data that has helped in understanding the evolutionary changes of the behavioral traits (witnessed in the gradual changes of formation of web architecture once initiated by the prehistoric ancestors of the extant spiders), against the backdrop of the past climatic conditions and ecosystems that the ancestors of the extant spiders had to acclimatized to (Robinson, 1975). Although morphological adaptations have played a key role in the alteration of behavioral patterns in acquiring genetic fitness, the behavioral traits have also played an important role in the regulation of reproductive success at the population level of the concerned spider. Whereas, environmental conditions have been attributed to the physical, physiological, biobehavioral, ecological, and genetic fitness of the species concerned (Vollarth and Selden, 2007). Thus, to define the evolution of spiders, a brief history of the origin and evolution of spiderwebs in terms of their formation and diversification needs to be elucidated.

Let's go back to the geological time scale to find the tentative time of the emergence of spiderwebs on our planet. Based on the outcome of their interdisciplinary research and investigations, paleontologists, arachnologists, evolutionary biologists and molecular phylogeneticists have ascertained that the "semi-terrestrial spiderlike arachnids," the progenitors of spiders and the maiden terrestrial carnivores on Earth—either intrigued by its prey-chasing behavior or driven by predatory attack aggravated their fleeing behavior—completed the transmigration from bodies of water to the terrestrial ecosystems likely between 419 and 358 million years ago (mya) during the Devonian Period (Shear et al., 1989; Selden et al., 1991). Over the course of this unique process of adaptation, the predecessors of these extant spiders went through stiff challenges to acclimatize to new climatic conditions and settle in new habitats as well as developing new locomotory tactics, new breathing techniques, unique reproductive measures, and improved hunting strategies. Parallels, silk-producing potential emerged along with its use in the evolution of web-spinning architecture in Araneae made it fit to evade and occupy entire inhabitable terrestrial ecosystems on the planet (Ronver, 1987; Selden and Edwards, 1989; Vollarth and Knight, 2005). Evolutionary biologists further theorized that parallel to the adaptive changes of physical, ecological, and physiological changes in the transmigration of arachnids from water to land, there was a unique behavioral adaptation (better be recognized as preadaptation) or behavioral preadaptive evolution in spiders as the formation of silk-producing glands built silk to use it for the next level of web-spinning needs that was likely initiated in the predecessors of extant spiders underwater and triggered the surge of diversification of arachnids throughout the terrestrial ecosystems as a result of intensive adaptive radiation (Decne, 1984; Rovner, 1987).

They postulated two hypothetical scenarios to define the origin and evolution of spider silk:

a. According to the first scenario, the spiders' eggs were wrapped in proteinaceous liquids exudate from the coxal glands adjacent to the abdominal legs. Eventually, those proteinaceous exudes dried to form fibrous sticky shields to protect the eggs from damage or from predatorily attacks (Shulz, 1987).

b. According to the second scenario, the lower abdomen of the spiders were likely to be smeared with proteinaceous mucous membrane that kept the gills moist in semiterrestrial condition before merging with book lungs found in extant spiders inhabiting terrestrial ecosystems (Damen et al., 2002).

Whether it fits in the first scenario (exudates) or the second (membranous smear), either form of structural entity had likely evolved with microfibrillary filaments that either acted as a water-retaining sac if the terrestrial habitat became dry and desiccated or like an air sac if the terrestrial habitat had temporarily submerged underwater. Scientists have observed further that in both scenarios, such saclike protective encasement resembled a "physical bubble," scientifically recognized as "Plastron lung," that probably coordinated exchange of gases from inside and outside, across the fibrous membrane (Messner and Adis, 1995). Going a step further, evolutionary biologists contemplated that the important role of such "hypothetical silken lung" would have rendered the predecessors of the extant spiders an evolutionary edge to acclimatize themselves for transitional adaptation before coming from the flood plains to settle in dried-out plains or terrestrial ecosystems (McQueen and McLay, 1983; Churchill and Raven, 1989). To substantiate such hypothetical scenarios, arachnologists have referred to some terrestrial spiders, such as Trapdoor spider *Idioctis* sp. and sheet weaver *Desis marina*, who sustained themselves in flood-prone zones for awhile by confined to their "Silken Sleeping Sacs" (Rovner, 1987). From an evolutionary perspective of spider silk, the two key integral parts are spinnerets and silk glands of which the spinnerets evolved from the hydraulic legs of the arachnids according to evolutionary biologists (Damen et al., 2002). Whereas the emergence of silk glands of the spiders remained in the labyrinth of hypothesis. A major group of arachnologists has theorized that silk is an excretory product, and during evolutionary change, the coxal glands were modified into silk glands of the extant spiders but the facts do not comply with the hypothesis, as coxal glands were determined to have originated from mesodermal tissues, whereas silk glands were of ectodermal origin (Kaston, 1964; Gertsch, 1979; Craig, 1997). During the period of transmigration, predecessors of the extant spiders were likely to be engaged in chasing and hunting potential prey, so self-protection was a serious concern in terms of enduring the scorching heat of the ecosystem during summers or tidal floods that turned arid land into the swamps that were significant impediments to existence. Thus, the evolution of silk webs seems to have primarily ensured the protection of spiders, their eggs, and spiderlings' survival. In the next level, the evolution of silk webs altered predator—prey dynamics as it restrained the spiders from roaming around to hunt potential prey and acquired neuro-sensory traits to trap a large number of prey by using its silk-web (Shear, 1986).

The arachnologists have gathered the fossil records of spider silk and silk webs from a number of primary and secondary sources. The meticulous studies on the fossil records on the formation of spider silk and spiderweb formation do not give any straightforward morphological information. However, it has helped scientists engage in a brief comparison of the structural formation of the ancestral representatives with their extant representatives that would definitely give the insight to understand the behavioral traits (silk production and web formation pattern) of the ancestors of the present-day spiders (Seilacher, 1967). Arachnologists have also mentioned the existence of trace fossils, such as silk threads, a fragment of webs, locomotion, and dwelling traces, or segments of silk-producing glands. These trace fossils have helped illuminate the progress of phylogenetic diversification of the spiders (Braddy, 1995; Zschokke, 2003; Gregory et al., 2006). Recently, they have done a comparative study on the locomotion traces of the mygalomorph spiders, the predecessors of the modern-day mygalomorph extant, traced from the 275-year-old Coconino Sandstone in Permian of the desert ecosystem of Arizona and New Mexico in the southwest United States of America to its descendants, such as *Aphonopelma chalcodes* or *Brachypelma* spp. Scientists have observed identical locomotory movements between the Permian predecessors in the desert ecosystems and their Anthropocene descendants, clearly establishing a phylogenetic line of descendant from the ancient spiders to their extant representatives (Sadler, 1993). A unique combination of burrows and spiderweb formation has also been studied that helped scientists understand the behavioral adaptation of spiders, such as sustaining themselves in the scorching heat of desert ecosystems. The availability of the trace fossils of *Seothyra* sp. from the Namibian desert in Miocene has given an impression to the arachnologists and evolutionary biologists that there was widespread distribution in the southern African desert regions (Dippenaar, 1990). The dwellings of the *Seothyra* sp. (which appeared like foot-impression of small antelope on the desert surface) have been carefully studied by scientists and they have noticed that this spider sheltered in vertical burrows having a horizontal web in the middle.

This behavioral evolution of building composite dwelling structures, such as "typical vertical burrow-horizontal web" likely ensured the survival of *Seothyra* sp. in the desert ecosystems while enduring scorching heat of arid and semiarid paleoclimate environment of West Africa for 80 million years (Schneider, 2004). The fossilized silk threads and the webs have also been recorded as "Bachofen-Echt Cretaceous Silk" entrapped in amber (Zschokke, 2003). The findings of entrapment of flies in amber and sections of silk-web have helped evolutionary biologists illuminate predator–prey dynamics of spiders in prehistoric time (Penalver et al., 2006).

According to the conventional phylogenetic investigations of any organism, the availability of mega-fossils and microfossils is likely a prerequisite criterion to compare the structural orientation (morphological and anatomical) of that prehistoric predecessor and its contemporary descendants, and extrapolation of such morpho-anatomical traits have often helped define the biobehavioral changes to find its phylogenetic root. But evolutionary biologists, paleontologists, geologists, and ecologists have also observed that paleoclimatic information of any geological regime is one of the key factors in understanding biobehavioral changes of the organism and further elucidating the biological evolution of the concerned organism. A majority of the arachnologists have been intrigued by the mysterious paleoclimatic changes of the Earth in the mid-Palaeozoic Era to understand the biobehavioral changes of spiders and to define their evolutionary progress. On the basis of a number of paleoclimatic reviews during the mid-Palaeozoic Era, most geologists and atmospheric scientists have argued that the oxygen level during this time was much less than 20%, the oxygen level in the current Anthropocene Era (Selden and Edwards, 1989). Going further, scientists argued further that in the low concentration of oxygen level during the mid-Palaeozoic Era, there was a lesser chance of formation of the ozone layer that forms a blanket around the Earth's atmosphere to stop deleterious and lethal radiation of UVB to reach the surface. Naturally, the formation and existence of any biological organism on Earth, above bodies of water, was considered theoretically impossible. The existence of life in bodies of water, during the mid-Palaeozoic was not be ruled out. Finally, experts on paleoatmospheric science ascertained that from 443.8 mya to 419.2 mya, during the Silurian Period, our planet was endowed with congenial levels of oxygen (almost the same level of oxygen concentration in today's atmosphere) which would allow the origin and evolution of biological entity on the terrestrial or semiterrestrial ecosystems on the planet (Ward et al., 2006). Although the critical examinations of fossils ascertained that the oxygen level on Earth during Upper Carboniferous (around 358.9 mya) was much higher than the contemporary levels of oxygen concentration (causing wildfires; resulting in the formation of heft charcoal as fossil fuel), in the Lower Carboniferous Period—around 298.9 mya, close to Devonian Period—there was a substantial drop in oxygen, leading to unavailability of terrestrial fossils. This paucity of fossils on Earth was paleontologically recognized as "Romer's Gap" (Ward et al., 2006). So an alteration of oxygen concentration during the Silurian-Carboniferous Period might have expedited transmigration for a number of animals including spiders, the progenitors of these extant animals, who likely had stepped from the water to invade and settle in terrestrial ecosystems. It is feasible to understand that a possible drop in oxygen level in the paleoatmospheric condition would render downgrading of respiratory organs (for optimization of breathing) or might accelerate the extent and multitude of extinction of the concerned animal that sustained such changes. At the same time, the surge in oxygen concentration in the palaeo-atmosphere might help the concerned organism to diversify and the upgradation of sudden organs might have been observed (e.g., the upgradation of the respiratory systems in spiders were witnessed with the formation of "book lungs with trachea") whenever it stepped into terrestrial ecosystems, as their ancestral lineages used to respire with tracheary gills in oxygenated water (Bromhall, 1987). As the lower level of oxygen and direct exposition of UVB radiation on Earth negated the origin and evolution of any biological entity, outside the bodies of water, a number of evolutionary biologists construed that a number of biological organisms (including spiders and amphibians) used to inhabiting aquatic ecosystems would venture on the land at night and take refuge in the clefts of rock crevices or burrows, silken canopy or spiderwebs or in the bodies of water (Vollarth and Selden, 2007). Even though the reflective potential of UVB radiation of the spiderwebs sounds hypothetical, arachnologists have observed that the silk threads, used in the spiderwebs of the ancestral lineages of the extant spiders, had a high resolution of the reflective potential of UVB irradiations that might have helped prehistoric spiders sustain themselves in adverse palaeo-environmental conditions (Craig, 2003). To obtain a unique portrayal of the history of origin, evolution, and diversification of spiders against the backdrop of paleoclimatic changes on Earth, let us take a peek through the geological timescale:

1. Ordovician spiders (485.4—443.8 million years ago)

Arthropods were the first group of species, to leave bodies of water and colonize terrestrial ecosystems around 450 mya. The earliest prehistoric arthropods probably evolved around 550 mya in the bodies of water (e.g., Parvancorina, Spriggina, etc.) (Swan, 2020). According to the phylogenetic classification, Chelicerata (includes spiders, scorpions, horseshoe crabs, mites, etc.)—considered a subgroup of arthropods—diverged around 445 mya. Chelicerae have two segments in their bodies (head and abdomen) and have been characterized by the presence of two "chelicerae" the pincers or fangs attached to the head used for hunting the prey. Paleontologists have recognized *Megarachne servienei* as the giant prehistoric spider (around 50 cm dia.), likely to have disappeared around 300 mya (Haug, 2020; Swan, 2020).

2. Silurian spiders (443.8–419.2 million years ago)

Trigonotarbida has been recognized as the earliest arachnids that looked like spiders but did not possess silk-producing glands. Evolutionary biologists predicted that the Trigonotarbida likely evolved between 420 and 290 mya (Swan, 2020). The Trigonotarbida comprise "octopedal" species of spiders, scorpions, mites, and ticks. The presence of a couple of chelicerae on its head segment appeared to be an additional pair of legs. The critical observation of evolutionary biologists has observed that the Trigonotarbida acquired a number of identical adaptations witnessed in modern spiders, such as possession of fine bristles that act as sense organs (sense of touch) and slit-sense organs that act like eardrum membrane that can detect vibrations and have the auditory ability. It also evolved compound eyes like human eyes (having cornea, retina, lens) which helped it inhabit and survive in different habitats in any congenial environment (Swan, 2020).

3. Devonian spiders (419.2–358.9 million years ago)

Atterocopus fimbriunguis, described for the first time by Shear et al. (1987), was discovered at the Brown Mountain fossil site in Gilboa, New York, and has been recognized as the oldest spider believed to have evolved in the mid-Devonian and is phylogenetically assigned to Trigonotarbida, an extinct group of arachnid, which likely emerged during the Palaeozoic Era (Selden et al., 1991). Based on the presence of spinnerets and other organs resembling spiders', a group of arachnologists promulgated that the *Atterocopus* spp. emerged with a set of plesiomorphic features that closely resembled those of *Liphistius* spp. (belonging to *Mesothele*, the primitive genus among the extant spiders) and has been recognized as a living fossil (Selden et al., 1991). Phylogenetically, in Araneae, *Atterocopus* spp. have been assigned as a sister group of all other spiders (Selden et al., 1991). To study the cuticular structure of the exterior part of *Atterocopus* spp., scientists have engaged in maceration of the cuticular fragments of this spider impregnated in the siltstone matrix and have used hydrofluoric acid for this maceration process (Selden, 1991). Scientists managed to salvage a section of the cuticular fragment of *Atterocopus* spp. and witnessed that it contained 19–20 bell-shaped spigots; the presence of such spigots (with a wide base and narrow shafts) showed structural resemblance to a number of spigots–bearing extant spiders of primitive origin. A group of arachnologists does not agree with the hypothetical explanation that *Liphistius* spp (belonging to mesothele) should be phylogenetically treated as a close ally of *Atterocopus* spp. as *Liphistius* spp. has a segmented abdomen, its single-articled median spinnerets possesses one spigot, whereas the lateral spinnerets are pseudosegmented, and these structural formations of spinnerets and spigots of *Liphistius* spp. are quite different from those of *Atterocopus* spp. (Bristowe, 1975; Haupt, 2003; Volarth and Selden, 2007). On the other hand, the critical structural investigations of the fossil of *Atterocopus* spp., revealed that its spigot possesses a fusiform spinneret resembling the mygalomorph spiders (the more recently evolved), characterized by the median spinnerets, the single-articled one possessing many spigots (Volarth and Selden, 2007). Hence, the extant representatives belonging to mygalomorph appeared to be close phylogenetically and descendants of *Atterocopus fimbriunguis*, the prehistoric, Devonian representative of arachnids, rather than considering the extant spiders belonging to mesothele.

However, the arachnologists have contended that spiders were one of the earliest terrestrial animals that emerged on Earth with unique behavioral adaptations to produce silk fibers, streaming out from its spigots, and these spigots were either attached to spinnerets or directly attached to the body surface. If we would like to inquire about the predator–prey dynamics of *Atterocopus fimbriunguis*, we should look into the faunal association of the ecosystem it inhabited in and scientists have noticed a rich association of amblypygids, arthropleurids, collembolans, diplopods, mites and myriapods, pseudoscorpions, scorpions, trigonotarbids and some chilopods (Shear and Selden, 2001). As far as the predator–prey relationship of the *Atterocopus fimbriunguis* is concerned, it is easier to identify collembolans, mites, and myriapods as its prey, whereas, the other arachnids such chilopods the have competed with *Atterocopus fimbriunguis* to gain their share of prey or extreme level of competition for prey might change their food habit and give rise to behavioral traits, such as cannibalism. Referring to the unique behavioral traits of some extant spiders who dwell in burrows with a trapdoor, evolutionary biologists have hypothesized that the biobehavioral features of the extant spider in the construction of burrows were likely passed down from their Devonian predecessors (Vollarth and Selden, 2007). Yet evolutionary biologists and arachnologists do not have a clear resolution to define the predator–prey dynamics as far as the emergence of prehistoric spiders are concerned in the Devonian Period as their findings did not support either coevolutionary emergence of prey insects of the Devonian predatory spiders or render any impression that adaptive radiation of insects was in full swing. The only fossil evidence of the insect *Rhyniognatha hirsti* was reported from the Rhynie Chert bed of the United Kingdom so far (Hirst and Maulik, 1926; Tillyard, 1928; Engel and Grimaldi, 2004). Surprisingly, the fossil existence of insects in the Lower Carboniferous Period did not present any sign of adaptive radiation this period was solely represented by *Delitzschala bitterfeldensis* found at a fossil site of Germany. However, this abruptly changed during the Upper Carboniferous Period, as entomologists witnessed a sudden spike in the population of insects in that period (Brauckmann and Schneider, 1996). The evidence of flying insects or small-sized flying insects were not there during the Devonian Period so origin and the biobehavioral evolution of the formation of silk threads and spiderwebs, which was assumed to be for entrapping insects was not supported logically (Eisner et al., 1964).

4. Carboniferous spiders (358.9–298.9 million years ago)

Around 30 fossil spider specimens have been reported from the Upper Carboniferous and most specimens belonged to Mesothelae, which gave an impression that the adaptive radiation of arachnids possibly happened in the geological regime belonging to Mesothelae. *Palaeothele montceauensis*, one of the prehistoric spiders, belonging to mesothele has been reported from a fossil site in France during the Upper Carboniferous (Selden, 1996). The presence of narrow-shaped sternum—a unique anatomical feature—found in *Palaeothele* spp. was possibly shared by its evolutionary descendants of extant mesothele spiders. The critical observation on the fossil specimens of Palaeothele has ascertained that it had five spinnerets, comprising one left anterior lateral (ALS), two anterior medians, one posterior lateral (PLS), and a fifth one adjacent to the PLS. It is assumed to be the substitute of posterior medial spinneret (PMS) (Selden, 1996). It has been firmly affirmed by arachnologists that the carboniferous spiders belonging to the family Arthromygalinidae and Arthrolycosidae, were of Mesothelae (Penny and Selden, 2006). Although, most arachnologists have recognized the spiders (e.g., *Archaeometa nephilina*, Eopholcus) as belonging to the family Arachaeometidae, have been identified as araneomorphs, have not been considered araneomorphs or spiders by the arachnid phylogeneticists (Penny and Selden, 2006).

The evolutionary history of insects was primarily ascertained to be around 310 mya, in the Upper Carboniferous Period, and the adaptive radiation of flying insects ensured the predominance of small-sized, flying fauna in the forest ecosystems. During this paleoclimatic regime, evolutionary diversification of spiderwebs was also witnessed in terms of design and architecture, but it was difficult to establish a coevolutionary relationship between these two factors that would help define predator—prey dynamics further. To establish a hypothetical link between these two issues, evolutionary biologists promulgated two scenarios:

1. For some reason, insects became involved in exploring the aerial ecosystem to chase airborne pollens, adapted biobehaviorally in the volant adaptation that indirectly triggered spiders to catch and consume them by forming sticky traps.
2. The hunting spree of the predatory spiders would force the insects to biobehaviorally adapt to fly. The substantiated the coevolutionary hypothesis of origin and evolution of wings in insects and of sticky, aerial spiderwebs.

The problematic part of substantiating such hypothesis is that there is a gross void of fossil evidence and it is difficult to establish a hypothesis merely based on a presumption about which circumstances led spiders to stop spinning webs near ground vegetation and instead to gradually climb; whether it was due to unavailability of prey at the ground level or would look for a different type of prey at the canopy level. However, the availability of fossil remnants of a diverse group of flying insects collected from the fossil site of Germany has confirmed that the adaptive radiation of insects and its evolution of volant adaptation likely took place in the Upper Carboniferous Era against the backdrop of emerging insect wings (Brauckmann et al., 1996).

Arachnologists have further observed that to maintain the predator—prey dynamics, the evolution of Devonian predators should not be considered the only factor, as the evolution of insect-eating trigonotarbids also took place during this paleoclimatic regime even though they were devoid of producing venoms and yielding silks. Based on the paleontological evidence of trigonotarbids in the Upper Palaeozoic Era, it clearly indicated that trigonotarbids might have had a greater influence in maintaining predator–prey dynamics of insects, rather than considering that predator–prey dynamics regulated by spider-insects dynamics. The ecologists have further observed that apart from trigonotarbids, the presence of a number of insect predators—amblypygids, chilopods, uropygids, scorpions, and some small vertebrates—in the terrestrial ecosystems during the Upper Carboniferous Period might have played a key role in maintaining the predator–prey dynamics of insects (Vollarth and Selden, 2007).

5. Permian spiders (298.9–251.9 million years ago)

Although plenty of insect fossils have been reported from the Permian Era, arachnologists were disappointed with the availability of arachnid fossils in the geological regime. After a long haul, paleontologists have come across the fossil remnants of *Aphantomartus*, the prehistoric arachnid, belonging to trigonotarbid (Rossler, 1998). *Arthrolycosa antiqua*, one of the extinct arachnids (belonging to subphylum Chelicerata), was reported from Mazon Creek, USA, and Kirov Oblast, Russia in the Upper Permian (estimated to have evolved around 300–250 mya) (Eskov and Selden, 2005). The paleontological remnants of *Permarachne novokshonovi*, one of the extinct arachnids, were discovered in the Koshelevka formation, from the Kungurian of Russia in the Upper Permian. It closely resembles another ancestral spider, Atterocopus, and has a structural resemblance with the modern spiders, except possessing a long taillike structure of scorpion (Selden and Eskov, 2005; Selden et al., 2008). By virtue of having plesiomorphies for spiders, the presence of a narrow sternum in Permarachne, highlighted its synapomorphic traits to its descendants of the extant spiders belonging to mesothele (Eskov and Selden, 2005). Critical observation of arachnologists has ascertained the presence of "flagelliform structures," protruding from its abdomen that has been considered "elongated spinnerets" in Permarachne (Eskov and Selden, 2005). A number of arachnologists have argued that elongated spinnerets were also noticed in the mygalomorph family Dipluridae and the araneomorph family Agellnidae but it was not found in any extant mesotheles (Vollarth and Selden, 2007). Such observations have rendered a group of scientists contemplate that there was possibly a great deal of diversity in mesothele during Upper Palaeozoic. Based on the presence of this kind of typical spinnerets, it has been further inferred that the evolution of funnel-web spiders took place in the Upper Palaeozoic Period as the funnel-webs were typically adapted to entrap jumping insects (Eskov and Selden, 2005).

The paleontologists have studied a diverse collection of fossils of flying insects and spiders from a few Fossil-Lagerstatten. It helped scientists understand that the Permian Period should be recognized as the regime of "rapid adaptive radiation" for flying insects and spiders that might have started an evolutionary arms race between them during this paleoclimatic regime around 270 mya (Vollarth and Selden, 2007). Naturally, evolutionary biologists expected that if spiders had gone through "rapid adaptive radiation," they would emerge as a diverse group of spiders. The emergence of morphological diversity of spiders might have been attributed to a diverse array of biobehavioral adaptations and the likely leading formation of a variety of aerial webs. But as a matter of fact, paleontological existence of such diverse type of aerial webs was missing and that gave the arachnologists an idea that the two mega-extinction events (end of Permian and end of Palaeozoic) which wiped out 82% genera of the Earth also wiped out the unique adaptations of those genera over the course of biological evolution (Erwin, 2006). Although it is difficult to assess the direct impact of mega-extinction events on spiders and insects, a sudden drop in the diversity of fossil history of spiders and insects in Lagerstatten (in the Permian deposits of the United States and Russia), had likely to be tumbled by the Permian-Triassic extinction.

6. Triassic spiders (251.90–201.36 million years ago)

Evolutionary biologists have observed that an extinction event (a tumble of its population) for any biological entity has always followed rapid adaptation that rendered a surge to its population; hence, it has often been recognized as a reset button for the initiation of the next level of evolution. So the process of natural extinction is considered an automated process to control population surges. Often it wipes out the existence of an unsuitable, extant lineage of the biological entity to pave the way for the emergence of a unique biological entity (which fits in such climate and habitat) to commence biological evolution. However, the Permian-Triassic extinction has ensured the disappearance of a diverse population of pterygote Palaeoptera and Archaeoptera and the vacant niche has been occupied by the new generation of insects—pterygote Neoptera. Entomologists have observed that the ancient population of Palaeoptera was not suitable for volant adaptations as they did not fold their wings, whereas the members of Neoptera could fold their wings and were capable of steady flight. It has been observed that the dragon and damselflies (Odonata) and the mayflies (Ephemereptera), the descendants of the Palaeoptera escaped the Permian–Triassic extinction and can be found in an extant state as living fossils.

The arachnologists have referred that the paleontological existence of *Rosamygale grauvogeli* (belonging to the family Hexathelidae) in an instance of evolution of the modern lineage of mygalomorphs (Selden and Gall, 1992). The name of the family of the spider Hexathelids implies that each individual possesses six spinnerets (ALS, PLS, and PMS), and the elongated PLS helps the spider spin a funnel web (e.g., Atrax). Arachnologists have noticed that the members of *Rosamygale* inhabited the burrows in the semi-arid region of deltas and trapped jumping insects by using its funnel web (Selden et al., 1999). The preponderance of the mygalomorphs in the fossil sites of the Triassic Era, was an important issue from a phylogenetic perspective, as it belongs to the sister group of infraorder araneomorphs (Selden et al., 1999). Paleontological studies on the ancient araneomorphs have revealed that they had a resemblance to the modern orb weavers, specifically, the members belonging to Araneoidea. Although most members belonging to Araneoidea have been recognized as the primary weavers of orb webs, yet some families and genera in this group spin webs secondarily (derived from orbs) (Coddington, 1986). Evolutionary biologists have observed that the origin and evolution of orb webs were initiated in the Triassic Era, likely to have taken place between 200 and 140 mya, as they were accustomed to climbing shrubs and trees during this time. Arachnologists further mentioned that the evolution of Triassic spiders belongs to mygalomorphs, araneomorphs, and the adaptive radiation of Neopteran pterygotes—the flying insects—has steered the predator–prey arms race to a new dimension.

7. Jurassic spiders (201.36–145 million years ago)

The first discovery of the fossil remnants of the spider in the Mesozoic Era has been represented by *Juraraneus rasnitsyni* (Eskov, 1984). Although a number of arachnologists have claimed that evolutionary history of arachnids in the Jurassic Period started with the discovery of *Jurarchaea zherikhini*, which has a striking structural resemblance to the extant araenomorphs (Eskov, 1987). The exploration of the Jiulongshan formation, an important fossil site in the Jurassic, has revealed the existence of a rich fossil-fauna of a diverse group of spiders belonging to mygalomorphs, araneoids, ulobortids, and palpimanoids (Vollarth and Selden, 2007). It must be mentioned here that the diversity of spiders in the Jurassic Period was not captured through the availability of its fossil representatives as the contemporary phylogenetic investigations revealed that the araneofauna in the Jurassic Period have been represented by the diverse range of spiders belonging to Mesothelae, Mygalomorphae, Hexathelidae, Araneomorphae, Araneoidea, Palpimanoidea, Uloboridae, etc. Careful studies of the araneofauna in the Jurassic Period have primarily ascertained the predominance of ancestral Orbiculariae (the orb-web weavers) in its distinct forms, the ecribellate araneoids and cribellate ulobordis. As the arachnologists have contended that the evolution of orb webs likely took place from the cribellate orbicuarians (cribellate uloborids), whereas the split of ecribellate orbicularians from the cribellate orbicularians likely took place during the Upper Jurassic Period or even earlier (Vollarth and Selden, 2007). Entomologists have observed that the adaptive radiation of hoometabolous insects took place during the Jurassic Period and that incident must have played a key role in maintaining predator–prey dynamics between the Jurassic insects and the spiders, further categorized under (1) burrow-dwelling spiders, (2) funnel-web spiders, and (3) orb-web spiders (Vollarth and Selden, 2007). The most interesting part of the evolutionary history of Jurassic spiders is the biobehavioral evolution of araneophagy or cannibalism (to chase and consume other spiders) among the Jurassic spiders that have identical biobehavioral traits in the extant members of Archaeidae. Consequently, the members of Archaeidae have elongated chelicerae identical to those used by their predecessors. The arachnologists have contended further that biobehavioral traits, such as the cursorial foraging adaptation, hunting of other spiders, and the emergence of araneophagy, are found in modern Archaeidae, which likely emerged from among its prehistoric ancestral lineages during the Jurassic Period.

8. Cretaceous spiders (145–66 million years ago)

Arachnologists and paleontologists have discovered tetragnathid and uloborid spider fossils from the fossil site of Cretaceous rocks in Spain. The researchers have critically observed that the typical tarsal claw pattern of these ancient Cretaceous spiders has a striking resemblance to the modern orb weavers (Selden, 1990; Selden and Penny, 2003). A fossil site for Cretaceous spiders in Spain was encountered as a lacustrine environment covered with volcanic ash. Scientists observed that there were plenty of tetragnathids around the lakeshore region in the evening for hunting plenty of insects. Scientists have studied the fossils deposited on the sedimentary bed of the lakes and observed that the ratio of fossil insects and the spiders was 1000:1 which gave them an idea that spiders did not fall into bodies of water like their counterparts did, prey insects found as fossils and abundantly in lacustrine environments. It was further assumed that the deposition of volcanic ashes on the spiderwebs led to their collapse that expedited the fall of spiders and prey insects. Eskov the Wunderlich (1994) have reported a number of spider fossils from Armenia and Azerbaijan. The arachnologists reported two juvenile specimens belonging to the Lagonomegopidae, a new arachnid family, preserved in amber deposit in Taimyr, Siberia, Russia from the Upper Cretaceous Period (Eskov and Wunderlich, 1994). Apart from these two juvenile samples, almost 50 fossil samples of the Upper Cretaceous spiders were reported from the same region belonging to Araneoidea, Dysderoidea, and Thomisoidea (Vollarth and Selden, 2007). The Cretaceous mygalomorphs (those were placed in the families of Antrodiaetidae, Atypidae, and Mecicobothriidae) were reported from the Lower Cretaceous Period in Mongolia and Siberia (Eskov and Zonshtein, 1990). The paleontologists and arachnologists observed that amber is an important fossil fixative that not only preserves the spiders but also preserves the silk threads and spiderwebs, so prehistoric amber deposits played an important role in understanding the diversity of spider fossils and forming an idea of the evolution and diversification of spiders in the Cretaceous Period (Zschokke, 2003). A brief account of the fossil spiders entrapped in different Cretaceous amber deposits are as follows:

1. The fossil spiders in amber deposits, belonging to Araneidae, Dictynidae, Lagonomegopidae, Linyphiidae, Oecobiidae, Oonopidae, and Sagestriidae, were reported from New Jersey, USA (Penney, 2002, 2004).
2. The fossil spiders in amber deposits belonging to Archaeidae, Pisauridae, Lagonomegopidae, and Ooniopidae were reported from Myanmar (Penney, 2003, 2004, 2005, 2006).
3. The fossil spiders in amber deposits belonging to Deinopidae and Linyphiidae have been reported from Lebanon (Penney, 2003; Penny and Selden, 2002).
4. The fossil spiders in amber deposits belonging to Lagonomegopidae, and Oonopidae, were reported from Canada (Penney, 2005, 2006).
5. The fossil spiders belonging to Nemesiidae, were reported from the Lower Cretaceous amber deposit in the Isle of Wight (Selden, 2002).
6. The fossil spiders belonging to Dipluridae were reported from the Cretaceous amber in Brazil (Selden et al., 2006).

From their origin, modern spiders have gone through the labyrinth of evolution for the last 250 million years, which has led to the diversification and sophistication of spiderwebs. Richard Dawkins, one of the eminent evolutionary biologists demonstrated that if the advancement of the biobehavioral traits of a spider has truly reflected in its gradual sophistication of web architecture, then it is the natural selection process that defined the functional ability of any web design (Duff, 2020).

References

Barth, F.G., 1985. Neurobiology of Arachnids. Springer-Verlag, Berlin viii±385 pp.

Braddy, S.J., 1995. The ichnotaxonomy of the invertebrate trackways of the Coconino Sandstone (Lower Permian), northern Arizona. N. M. Mus. Nat. Hist. Sci. Bull. 6, 219–224.

Brauckmann, C., Brauckmann, B., Gröning, E., 1996. The stratigraphical position of the oldest known Pterygota (Insecta, Carboniferous, Namurian). Ann. Soc. Geol. Belg. 117, 47–56.

Brauckmann, C., Schneider, J., 1996. Ein unter-karbonisches Insekt aus dem Raum Bitterfeld/Delitzsch (Pterygota, Arnsbergium, Deutschland). Neues Jahrbuch Geol. Palaontol. Monatsh. 17–30.

Bristowe, W.S., 1958. The world of spiders. In: New Nat. Ser. Collins, London viii+304 pp.

Bristowe, W.S., 1975. A family of living fossil spiders. Endeavour 34, 115–117.

Bromhall, C., 1987. Spider tracheal systems. Tissue Cell 19, 793–807.

Churchill, T.B., Raven, R.J., 1989. The circumtropical distribution and habits of the intertidal mygalomorph spider Idioctis (Barychelidae). Rep. Dep. Biol. Univ. Turku 15.

Coddington, J.A., 1986. The Monophyletic Origin of the Orb Web, pp. 319–363. See Shear 1986.

Coddington, J.A., Levi, H.W., 1991. Systematics and evolution of spiders (Araneae). Annu. Rev. Ecol. Systemat. 22, 565–592.

Craig, C.L., 1997. Evolution of arthropod silks. Annu. Rev. Entomol. 42, 231–267.

Craig, C.L., 2003. Spiderwebs and Silks: Tracing Evolution from Molecules to Genes to Phenotypes. Oxford Univ. Press, New York viii+200 pp.

Damen, W.G.M., Saridaki, T., Averof, M., 2002. Diverse adaptations of an ancestral gill: a common evolutionary origin for wings, breathing organs, and spinnerets. Curr. Biol. 12, 1711–1716.

Decae, A.E., 1984. A theory on the origin of spiders and the primitive function of spider silk. J. Arachnol. 12, 21–28.

Dippenaar, A.S., 1990. A revision of the African spider genus Seothyra Purcell (Araneae: Eresidae). Cimbebasia 12, 135–160.

Duff, S., 2020. Richard Dawkins Ridiculed on Twitter for Baffling Question about Spiders and Lions. https://www.mirror.co.uk/3am/celebrity-news/richard- dawkins-gets-rinsed-twitter-23231087.

Eberhard, W.G., 1981. Construction behaviour and the distribution of tensions in orb webs. Bull. Br. Arachnol. Soc. 5, 189–204.

Eisner, T., Alsop, R., Ettershank, G., 1964. Adhesiveness of spider silk. Science 146, 1058–1061.

Engel, M.S., Grimaldi, D.A., 2004. New light shed on the oldest insect. Nature 427, 627–630.

Erwin, D.H., 2006. Extinction. How Life on Earth Nearly Ended 250 Million Years Ago. Princeton Univ. Press, Princeton, NJ viii+296 pp.

Eskov, K.Y., 1984. A new fossil spider family from the Jurassic of Transbaikalia from (Araneae: Chelicerata). Neues Jahrbuch Geol. Palaontol. Monatsh. 645–653.
Eskov, K.Y., 1987. A new archaeid spider (Chelicerata: Araneae) from the Jurassic of Kazakhstan, with notes on the so-called "Gondwanan" ranges of recent taxa. Neues Jahrbuch Geol. Palaontol. Abhand. 175, 81–106.
Eskov, K.Y., Selden, P.A., 2005. First record of spiders from the Permian period (Araneae: Mesothelae). Bull. Br. Arachnol. Soc. 13, 111–116.
Eskov, K.Y., Wunderlich, J., 1994. On the spiders of Taimyr ambers, Siberia, with the description of a new family and with general notes on the spiders from the Cretaceous resins (Arachnida: Araneae). Beitraege Araneol. 4, 95–107.
Eskov, K.Y., Zonshtein, S., 1990. First mesozoic mygalomorph spiders from the Lower Cretaceous of Siberia and Mongolia, with notes on the system and evolution of the infra-order Mygalomorphae (Chelicerata: Araneae). Neues Jahrbuch Geol. Palaontol. Abhand. 178, 325–368.
Foelix, R.F., 1996. Biology of Spiders, second ed. Oxford Univ. Press, Oxford. viii+200 pp.
Gertsch, W.J., 1979. American Spiders. Van Nostrand Reinhold, New York.
Gregory, M.R., Campbell, K.A., Zuraida, R., Martin, A.J., 2006. Plant traces resembling Skolithos. Ichnos 13, 205–216.
Haug, C., Reumer, J.W.F., Haug, J.T., et al., 2020. Comment on the letter of the Society of Vertebrate Paleontology (SVP) dated April 21, 2020 regarding "Fossils from conflict zones and reproducibility of fossil-based scientific data": the importance of private collections. PalZ 94, 413–429. https://doi.org/10.1007/s12542-020-00522-x.
Haupt, J., 2003. The Mesothelae—a monograph of an exceptional group of orders (Araneae: Mesothelae). Zoologica 154, 1–102.
Hirst, S., Maulik, S., 1926. On some arthropod remains from the Rhynie chert (Old Red Sandstone). Geol. Mag. 63, 69–71.
Kaston, B.J., 1964. The evolution of spider webs. Am. Zool. 4, 191–207.
Krink, T., Vollrath, F., 1997. Analysing spider web-building behaviour with rule-based simulations and genetic algorithms. J. Theor. Biol. 185, 321–331.
McQueen, D.J., McLay, C.L., 1983. How does the intertidal spider Desis marina (Hector) remain underwater for such a long time? N. Z. J. Zool 10, 383–391.
Messner, B., Adis, J., 1995. There is only facultative plastron respiration in diving web spiders (Araneae). Dtsch. Entomol. Z. 42, 453–459.
Nentwig, W., 1986. Ecophysiology of Spiders. Springer-Verlag, Berlin.
Opell, B.D., 1989. Functional associations between the cribellum spinning plate and capture threads of Miagrammopes animotus (araneida, uloboridae). Zoomorphology 108, 263–267.
Opell, B.D., Bond, J.E., 2001. Changes in the mechanical properties of threads and the evolution of modern orb-weaving spiders. Evol. Ecol. Res. 3, 567–581.
Peñalver, E., Grimaldi, D.A., Delclòs, X., 2006. Early Cretaceous spider web with its prey. Science 312, 7161.
Penney, D., 2002. Spiders in upper cretaceous amber from New Jersey (arthropoda, araneae). Palaeontology 45, 709–724.
Penney, D., 2003a. Afrarchaea grimaldii, a new species of Archaeidae (Araneae) in Cretaceous Burmese amber. J. Arachnol. 31, 122–130.
Penney, D., 2003b. A new deinopoid spider from Cretaceous Lebanese amber. Acta Palaeontol. Pol. 48, 569–574.
Penney, D., 2004a. New spiders in upper cretaceous amber from New Jersey in the American museum of natural history (arthropoda, araneae). Palaeontology 47, 367–375.
Penney, D., 2004b. A new genus and species of Pisauridae (Araneae) in Cretaceous Burmese amber. J. Syst. Palaeontol. 2, 141–145.

Penney, D., 2005. The fossil spider family Lagonomegopidae in Cretaceous ambers with description of a new genus and species from Myanmar. J. Arachnol. 33, 439–444.
Penney, D., 2006. Fossil oonopid spiders in Cretaceous ambers from Canada and Myanmar. Palaeontology 49, 229–235.
Penney, D., Selden, P.A., 2002. The oldest linyphiid spider, in lower cretaceous Lebanese amber (araneae, linyphiidae, linyphiinae). J. Arachnol. 30, 487–493.
Penney, D., Selden, P.A., 2006. Assembling the Tree of Life—phylogeny of Spiders: a review of the strictly fossil spider families. In: Deltshev, C., Stoev, P. (Eds.), European Arachnology 2005, Acta Zool. Bulg. Suppl., vol. 1, pp. 25–39.
Robinson, M.H., 1975. The evolution of predatory behaviour in araneid spiders. In: Baerends, G., Beer, C., Manning, A. (Eds.), Function and Evolution in Behaviour. Clarendon, Oxford, United Kingdom, pp. 292–312.
Rössler, R., 1998. Arachniden-Neufunde im mitteleuropäischen Unterkarbon bis Perm—Beitrag zur Revision der Familie Aphantomartidae Petrunkevitch 1945 (Arachnida, Trigonotarbida). Paläontol. Z. 72, 67–88.
Rovner, J.S., 1987. Nests of terrestrial spiders maintain a physical gill: flooding and evolution of silk constructions. J. Arachnol. 14, 327–337.
Sadler, C.J., 1993. Arthropod trace fossils from the permian de chelly Sandstone, northeastern Arizona. J. Paleontol. 67, 240–249.
Schneider, G., 2004 (Samml. Geol. Führer 97). In: The Roadside Geology of Namibia. Gebrüder Borntraeger, Stuttgart, 294 pp.
Seilacher, A., 1967. Fossil behaviour. Sci. Am. 217, 72–80.
Selden, P.A., 1990. Lower cretaceous spiders from the Sierra de Montsech, north-east Spain. Palaeontology 33, 257–285.
Selden, P.A., 1996. First fossil mesothele spider, from the Carboniferous of France. Rev. Suisse Zool. 2, 585–596.
Selden, P.A., 2002. Missing links between ARGYRONETA and CYBAEIDAE revealed by fossil spiders. J. Arachnol. 30 (2), 189–200.
Selden, P.A., Anderson, H.M., Anderson, J.M., Fraser, N.C., 1999. The oldest araneomorph spiders, from the Triassic of South Africa and Virginia. J. Arachnol. 27, 401–414.
Selden, P.A., Casado, F. da C., Mesquita, M.V., 2006. Mygalomorph spiders (Araneae: Dipluridae) from the lower cretaceous crato lagerstätte, araripe basin, north-east Brazil. Palaeontology 49, 817–826.
Selden, P.A., Edwards, D., 1989. Colonisation of the land. In: Allen, K.C., Briggs, D.E.G. (Eds.), Evolution and the Fossil Record. Belhaven, London, pp. 122–152 xiii + 265 pp.
Selden, P.A., Gall, J.-C., 1992. A Triassic mygalomorph spider from the northern Vosges, France. Palaeontology 35, 211–235.
Selden, P.A., Penney, D., 2003. Lower cretaceous spiders (arthropoda: arachnida: Araneae) from Spain. Neues Jahrbuch Geol. Palaontol. Monatsh. 175–192.
Selden, P.A., Shear, W.A., Bonamo, P.M., 1991. A spider and other arachnids from the Devonian of New York, and reinterpretations of Devonian Araneae. Palaeontology 34, 241–281.
Selden, P.A., Shear, W.A., Sutton, M.D., 1989. Fossil evidence for the origin of spider spinnerets, and a proposed arachnid order. Proc Natl Acad Sci U S A. 105 (52), 20781–20785. https://doi.org/10.1073/pnas.0809174106. Epub 2008 Dec 22.
Shear, W.A., 1986. Spiders: Webs, Behavior, and Evolution. Stanford Univ. Press, Stanford xiii + 492 pp.
Shear, W.A., Palmer, J.M., Coddington, J.A., Bonamo, P.M., 1989. A Devonian spinneret: early evidence of spiders and silk use. Science 246, 479–481.

Shear, W.A., Selden, P.A., 2001. Rustling in the undergrowth: animals in early terrestrial ecosystems. In: Gensel, P.G., Edwards, D. (Eds.), Plants Invade the Land: Evolutionary and Environmental Perspective. Columbia Univ. Press, New York, pp. 29–51 x + 304 pp.

Shear, W.A., Selden, P.A., Rolfe, W.D.I., Bonamo, P.M., Grierson, J.D., 1987. New terrestrial arachnids from the Devonian of Gilboa, New York (arachnida: trigonotarbida). Am. Mus. Novit. 2901, 1–74.

Shultz, J.W., 1987. The origin of the spinning apparatus in spiders. Biol. Rev. 62, 89–113.

Snodgrass, R.E., 1952. A Textbook of Arthropod Anatomy. Cornell Univ. Press, Ithaca, NY.

Swan, T., 2020. The Evolution of Spiders and Their Remarkable Adaptations. https://owlcation.com/stem/Spider-Evolution-Adaptations-and-Ancestors.

Tillyard, R.J., 1928. Some remarks on the Devonian fossil insects from the Rhynie chert beds, old red Sandstone. Trans. R. Entomol. Soc. Lond. 76, 65–71.

Townley, M., Tillinghast, E., Cherim, N., 1993. Moult-related changes in ampullate silk gland morphology and usage in the araneid spider Araneus cavaticus. Philos. Trans. R. Soc. Lond. Ser. B 340, 25–38.

Vollrath, F., 1987. Altered geometry of web in spiders with regenerated legs. Nature 328, 247–248.

Vollrath, F., 1992. Analysis and interpretation of orb spider exploration and web-building behaviour. Adv. Stud. Behav. 21, 147–199.

Vollrath, F., Knight, D., 2005. Biology and technology of silk production. In: Steinbuchel, A., Doi, Y. (Eds.), Biotechnology of Biopolymers: From Synthesis to Patents, vol. 2. Wiley-VCH, Weinheim, pp. 873–894.

Vollarth, F., Selden, P.A., 2007. The role of behavior in the evolution of spiders, silks, and webs. Annu. Rev. Ecol. Evol. Syst. 38, 819–846.

Ward, P., Labandeira, C., Laurin, M., Berner, R.A., 2006. Confirmation of Romer's Gap as a low oxygen interval constraining the timing of initial arthropod and vertebrate terrestrialization. Proc. Natl. Acad. Sci. U. S. A. 103, 16818–16822.

Witt, P.N., Reed, C.F., Peakall, D.B., 1968. A Spider's Web: Problems in Regulatory Biology. Springer-Verlag, Berlin viii + 107 pp.

Zschokke, S., 2003. Spider-web silk from the early cretaceous. Nature 424, 636–637.

CHAPTER ELEVEN

Impact of environmental parameters on the biodiversity, web formation, and niche properties of spiders

Abstract

Arachnologists, evolutionary biologists, and environmental biologists have meticulously studied the alteration of environmental conditions and direct impact of pollutants on the metabolic alterations of a large number of spiders with diversified modes of hunting strategies, behaviors, distinct shapes and sizes, well-diversified ecophysiological traits, and web-construction patterns belonging to different taxonomic subcategories. Environmental scientists have noticed that spiderwebs are one of the most effective nonselective accumulators of air pollutants, which has helped scientists in long-term monitoring of the environmental conditions of microenvironments. Scientists have studied the alteration of metabolic conditions in relation to the exposition of different environmental pollutants to understand the state of ecophysiological traits at the species level to interpret the results in terms of impacts on environmental changes in the species and ecodiversities of spiders and the state of their microevolutionary changes against the backdrop of ever-changing environmental conditions.

Keywords: Ecophysiological changes; Environmental pollution; Microevolution; Spiderwebs.

To reveal key environmental factors that affect spider communities, abiotic and biotic habitat parameters (e.g., temperature, air humidity, soil characteristics, and vegetation parameters) were systematically assessed around each pitfall trap. Spider species showed a small-scale distribution pattern on the forest floor, significantly affected by litter type, degree of canopy closure, temperature, and humidity parameters, as well as the cover of grass, moss, herb, and natural regeneration according to discriminant analysis and ordinations … Nevertheless, the high similarity in the composition of spider assemblages from different stand types but with similar formations of ground vegetation, microclimatic parameter, and canopy closure displayed a considerable impact of environmental factors not necessarily bound to the stand type itself.

T. M. Ziesche and M. Roth

Spider Evolution
ISBN: 978-0-323-90041-6
https://doi.org/10.1016/B978-0-323-90041-6.00002-4

For some time, arachnologists and interdisciplinary scientists have considered spiders the best indicators of environmental qualities, as qualitative synecological investigations revealed that influential climatic and environmental factors maintain the dynamics of the vegetational heterogeneity of habitats. Such uninterrupted changes in the assemblage composition of habitat biota at local levels have typically maintained the species or alpha diversity of particular species in the concerned biota (in a site habitat at the local level) in a mode wherein changes occur against a backdrop of changing environmental variables (Whittaker, 1960). Going a step further down the trail of appraising species biodiversity against a landscape of species diversity at the habitat level and differentiation among habitats, diversity represents "absolute species turnover" refereed by the ratio of the local to regional diversity of the concerned species (Whittaker, 1960). From an ecological perspective, the beta diversity of any species (e.g., arachnid species such as the modern spider) has already revealed scenarios such as "spatial replacement in species composition," which is the cardinal issue in understanding the status of concerned species in "local communities assemblage" related to the concerned "space" going through several changes because of environmental or geographical factors (geographic distance, accessibility, etc.) or a combination of both (Rodriguez-Artigas et al., 2016). In contemporary investigations of the impact of environmental variables on the alpha and beta diversity of spider communities at the regional level in the Campos and Malezales ecoregions (Corrientes), Argentina, Sandra M. Rodriguez-Artigas and her research colleagues of the IEBI-Facultad de Ciencias Naturales, Universidad Nacional de Salta, Argentina, coordinated an experimental protocol wherein they used spiders as model taxa to witness (Rodriguez-Artigas et al., 2016).

1. Whether there is a declining rate of similarity that corresponds to the geographical distance between the Campos and Malezales ecoregions of Argentina, which are geographically 13 km apart;
2. Whether there is any influence in climatic or environmental factors on the assemblage composition of spider families at the regional level; and
3. Whether the differential dispersal potential of spider families found in qualitative experimental observation indicates any cardinal factor that influences its community structure at a regional level.

To coordinate such an experimental format, the spider samples were collected with a G-Vac from vegetation of the six experimental locations of the grassland ecosystems of the Campos and Malezales ecoregions, Argentina (Rodriguez-Artigas et al., 2016). The research team under the mentorship of Rodriguez-Artigas performed statistical analysis that primarily helps in achieving the preliminary objective of evaluating the alpha diversity of spiders by coordinating their regional and site-specific inventories. The preliminary observation of this experiment is unique, as it reveals that the similarity between the spider communities is declining because the geographic distance between the adjacent ecoregions has increased (Soininen et al., 2007). Scientists have observed that the influence of abiotic factors, interaction with adjacent biotic factors, environmental variables, and the occurrence of stochastic events play important roles in altering the assemblage of the community structure of spiders. Moreover, it has been observed that differential dispersal potential among different spider groups intervenes in the formation of the regional-level community structure of spiders (Rodriguez-Artigas et al., 2016). Furthermore, it helps in evaluating the impact of different environmental variables on the beta diversity of the concerned spider communities at the regional level and key integral observations (functional variables that trigger the beta diversity of any organism) such as species replacement, nesting behavior, etc. Careful observation of the experimental data analyzed in this experimental format along with previous investigations revealed several unique results (Sunderland and Samu, 2000; Wise, 2006; Simo et al., 2011; Viera, 2011; Ferro and Romanowski, 2012; Rodriguez-Artigas et al., 2016):

1. The extent and magnitude of the regional beta diversity of any species are higher if the species is of high to intermediate vagility potential (species that are promigratory, have extensive dispersal ability, and have high adaptive potential), whereas nesting potential has normally been observed in spiders with low dispersion as well as members with low vagility potential.
2. This experiment has affirmed that the species with higher dispersal potential reveal great species replacement potential, the geographical discontinuity of the grasslands between Campos and Malezales is believed to stall the transitional dispersion of spiders across the geographical barrier, and microhabitats act as refuges adjacent to the barrier between the two ecoregions.
3. Dispersal potential, which is less prevalent in spiders with less vagile potential, has been influenced primarily by the geographical distance between the two adjacent ecoregions and biotic factors such as inter- and intraspecific competition.

In forest ecosystems, spiders are normally found living in tree cavities, and the intervening factors that impact the formation of the community structure of their microhabitats, such as nests in tree cavities, also depend on the nature of the bark of the trees, seasonal changes, intraguild predation, etc. (Černecká et al., 2017). A group of scientists under the mentorship of Ľudmila Černecká, Institute of Forest Ecology SAS, Slovakia, engaged in the investigation of spider abundance along with the taxonomic and functional composition of spiders in nest boxes in two mixed forest stands in Central Slovakia between 2012 and 2013 for three consecutive seasons. The experiments were performed with 1211 spider samples belonging to 31 species, with the samples taken from 60 experimental nest boxes. The preliminary results of this experiment were quite interesting, as the investigating arachnologists observed that the presence of insectivorous birds could reduce 67% of the gross abundance of the spider community in a region, and web spiders were the main victims of bird predators; drastic shrinkage in their population size might alter the dynamics of the guild composition of the trophic dynamics of the concerned ecosystems. The key observations of those investigations were quite interesting (Černecká et al., 2017):

1. Spider abundance surges in autumn, as entire spider communities are engaged in securing safe hideouts for winter.
2. The seasonal dominance of Anyphaenidae, Clubionidae, and Philodromidae (technically recognized as "other hunters") in the concerned spider communities are observed in spring and autumn, whereas the dominance of Linyphiidae (recognized as "sheet web weavers") occurs in summer.

Černecká and her fellow research investigators observed that the activated state and intraguild predation behavior of the small-sized winter-active spiders (active in the autumn and winter seasons) play a cardinal role in maintaining the equilibrium of the population dynamics of spider communities at the local level (Černecká et al., 2017).

3. It was also noticed by the Černecká (2017) team that the morphological pattern of the bark of the trees affects the community structure of spiders, as the dominance of "space web weavers" has been observed in trees with rough bark over trees with smooth bark. The extent of intraguild predation by *Anyphaena accentuata* in philodromids has declined more substantially in rough-bark tree habitats than in smooth-bark tree habitats (Melo-Braga et al., 2020).

The foregoing experiment proves that the interactions of the biotic (alteration of predator—prey dynamics by switching from monophagy to polyphagy or vice versa) and abiotic factors (unusual environmental changes like prolonged drought or a long winter season) affected the formation of the community structure of spiders, and gradual changes in these factors influenced the alteration of functional traits of spiders living in wild habitats. Meticulous observations by arachnologists and ecologists have revealed several key biotic factors controlling the population dynamics of spiders in the diverse array of their habitats, ranging from wet meadows to forests and agricultural fields within polyphagy, restricted nutritional selections, a high level of hunting potential, etc. (Wilczek et al., 2017). Careful experimental observations by arachnologists and field zoologists have revealed that anthropogenic changes in habitats, such as conversion of wild habitats to pastures, farmland, agricultural fields, mining operations, heavy industrial operations, etc., have led to changes in wild habitats being forced to sustain excessive levels of environmental pollution, absorb heavy metals and toxic chemicals, and adjacent water bodies polluted with effluent treated sludge, etc., which has altered the ecological dynamics of ecosystems. Naturally, the act of predation and the hunting efficiency of small stakeholders (insects, spiders, etc.), specifically the obligatory predators of the concerned ecosystems, have been challenged, as the tolerance level of acclimatizing to highly polluted conditions varies among different organisms under various environmental conditions, which further depends on the xenobiotic[1]—metabolic potential of the concerned organisms for pollutants infused in organisms via either "ingestion or contact" (Wilczek et al., 2017). Multitasking investigations to coordinate any experimental protocol to assess a series of physiological investigations on spiders at the molecular level are very complicated. These include enzymatic and nonenzymatic detoxification indicators, the impact of heat shock proteins, assessing the cell death process, and xenobiotic changes in the cellular level among various spider species. Nevertheless, molecular biologists and invertebrate physiologists have observed that the impact of environmental pollutants and toxins on spiders is species- and sex-specific (Wilczek et al., 2017). The qualitative appraisal of environmental stressors on spiders reveals that the "stressing factors may allow identification of defense strategies triggered in spiders by different environmental stressors" that have played a key role in determining the survival ability of certain species over others in adapting to new ecological

[1] Xenobiotic: It has also been recognized as xenobiotics. The xenobiotics are neither found in the organism nor the environment. Scientists observed that some of the naturally occurring chemicals or endobiotics become xenobiotics due to the presence of it in nature at a very high level of concentrations.

niches and recolonizing the anthropogenically maneuvered habitat and appearing to occupy the diversified ecological niches that have gone through the intermittent changes of an environmental roller coaster during the regime of Anthropocene. From this standpoint, a clear delineation is needed between the two closely related terms—i.e., habitat and niche—when the structural, behavioral, and functional identity of any species is concerned from an ecological perspective. We designate any space with a set of environmental conditions that support the concerned biological species in adapting to live normally a "habitat," whereas the functional role of any species that helps it survive in its environment—diet, shelter, etc.—is called a "niche." From an ecological perspective, the niche is related to energy production and its flow through the different trophic levels in an ecosystem.

In 2015, Yi Qun Chai and his research colleague Dustin Wilgers at the Natural Science Department, McPherson College, Kansas, USA, shared their research findings on the adaptive changes of many spider species in a certain habitat that sustained annual cycles of environmental swings. Chai and Wilgers (2015), particularly the behavioral and physiological adaptations among several diverse groups of spiders to remain alive under variable environmental conditions. It was generally observed that the formation of a lower group of invertebrate communities and dominance of certain species over others in those communities, such as "short-lived arthropods," is influenced by the diverse nature of habitats, as the environmental variables in such microenvironments around specific habitats, such as temperature, humidity, availability of light, and the presence of potential predators, become changed at the temporal level (in years, months, hours, minutes) (Foelex, 1996). It is conceivable that surviving under undulating environmental conditions would result in better adaptability of spider communities living in such environments than those spider communities living under apparently static environmental conditions, as spider fitness is primarily affected. It has been affirmed by arachnologists that continuous environmental undulations supposedly promote plastic traits (e.g., behaviors), to favor "individuals to respond consistently to their inconsistent environment" (Primer, 2004). The arachnid scientist duo observed that the wolf spider generally uses a diverse range of refuges (burrows) to escape from its den at the surface level compared with those with burrows, because burrowers usually abandon the cave to avoid many challenging issues, broadly categorized as (Chai and Wilgers, 2015).

1. Biotic factors: the presence and absence of potential predators around the burrows and the variation of foraging potential against the backdrop of the rotation of the seasonal cycle;
2. Abiotic factors: changes in environmental variables such as light, temperature, humidity, and wind current; or
3. A combination of both factors.

Chai and Wilgers (2015) coordinated an experiment on the effect of light exposure and changing temperature conditions on the refuge of the Wolf spider *Rabidosa punctulata,* and they choose 4.4°C as cold and 26.7°C as hot and set the optimization of light by standardizing the (light vs. dark in N = 20/treatment) light exposure. Scientists had observed that changes in light exposure and temperature in the burrows (refugia) of spiders played a key role in their functional modes, though changes in the functional behavior of spiders vary from species to species. From the general perspective of spider biology, scientists have concluded that most spiders, specifically wolf spiders, are active in warm and dark conditions and remain mostly engaged in foraging at night, whereas during the cold or winter season, most spiders take refuge inside the burrows, and those spiders can hardly utilize their inner energy in the course of metabolism to adjust to any abrupt changes in environmental variables during a brief temporal cycle like endotherms. (Chai and Wilgers, 2015).

In 2012, an investigation was conducted on Arctic spider species' abundance and distribution in Svalbard, Norway, where arachnologists dealt with 15 native spider species belonging to Linyphiidae and Gnaphosidae. The experiment took place at four experimental sites at around a 300 m (above sea level) altitudinal location of Spitsbergen, Svalbard (Dahl et al., 2018). At the end of the experiment, scientists observed that only 2 (*Erigone arctica palearctica* and *Hilaria gracilis*) of the 15 spider species were found at all four experimental sites, and those 2 species represented 54% of the entire spider population. Careful analysis of this experimental results (in accord with the generalized linear model protocol) helped scientists determine that the distribution of spiders in the Arctic region was more integrated to topographical factors and vegetational heterogeneity than key environmental factors such as temperature (Dahl et al., 2018). In a unique contemporary investigation done by Marzena Stańska, Faculty of Natural Science, Siedlce University of Natural Sciences and Humanities, Poland, along with her fellow researchers in the Bug River Valley Grasslands (in Eastern Poland), scientists observed the impact of habitat complexity on the size of the spider body, and in this experiment, they targeted two species of *Alopecosa* spiders, *Alopecosa cuneata* and *A. pulverulenta* (of Lycosidae). Several ecological investigations revealed that vegetational heterogeneity and complex nature of habitat impacted the dynamics of taxonomic composition and diversity of spiders and a number of individuals in a population and its assemblages at local and regional levels but it remains unknown whether alteration of such factors would influence changes in the size of the body of the concerned spiders' species. To find the indirect impact of environmental factors (such as the type of habitat and its complexity) on the body sizes of the two species of the genus *Alopecosa*, Stańska and her fellow researchers conducted unique ecological experiments in 12 experimental plots of Bug River Valley of three distinct types:

a. Mesic meadows
b. Sandy grassland, and
c. Xerothermic grassland

The spider samples were collected by pitfall traps from three distinct types of habitats of the experimental plots from the spring to autumn season. Random sampling of spiders trapped by the pitfall traps revealed 315 males and 104 females belonging to *A. pulverulenta* and 334 male and 168 females of *A. cuneate* (Stańska et al., 2018). Based on statistical analysis of the experimental data (following the generalized linear mixed model), Stańska and her colleagues concluded that habitat type did not impact spider body size, but rather, formation of a larger body had a collinear relationship with habitat complexity. From this experiment, scientists have contemplated that larger spider body size is directly related (a collinear relationship) to the enhanced level of complexity of the habitat, where predator pressures are substantially less (Stańska et al., 2018). In their experimental observations, Stańska and her fellow researchers observed that carnivorous birds were the key spider predators, so a valid hypothesis is that if a predator engaged in hunting bigger-sized prey in an open environment rather than chasing smaller or tiny groups of prey (spiders), the pressure of selective predation would act as a natural downsizing method in the open habitat, supposedly keeping large-sized spiders at low levels in heterogeneous spider communities. Hence, scientists engaged in field-based ecological investigation based on spider assemblage and composition in the Bug River Valley Grassland hypothesized that large-sized spiders would take refuge in a dense type of habitat covered with complex vegetation to survive predatory pressure from carnivorous birds. Stańska and colleagues further concluded that habitat complexity, as a key environmental (indirect) factor, acts as a selection pressure in the assemblage of larger-sized spiders as well as the main inducing factor of assemblage of spiders according to their body size in any habitat or ecological niche (Stańska et al., 2018).

From an evolutionary perspective, when ecologists and arachnologists have dealt with the widespread colonization of the terrestrial habitat developed by the small arthropods like spiders, they have identified three key issues that affect the adaptation process, and these issues are as follows:

a. The small body size of spiders is a real challenge to deal with in the lower index of its large surface/volume ratio, which downgrades the energy exchange and maintenance of energy dynamics integrated to the energy metabolism of the concerned species, as an adaptation of any species in an environment seems to be difficult if it is evolved with an inherent lower rate of metabolism.

b. Terrestrial habitats comprise a diverse range of habitats with a diverse array of edaphic, topographic, climatic, and ecological variations, with substantial fluctuations in temperature and humidity (e.g., desert, arctic/montane, and savannah grass ecosystems). It is thus difficult for tiny, prehistoric, invertebrate creatures like spiders to go through the process of adaptation with primitive physical, physiological, anatomical features for coping with such environmental shifts in a way that could render the concerned species sustainable under an ecoclimatic condition.

c. The internal or endogenous concentration of water in a spider is higher than the external concentration of water in its environment, which renders it physiologically disadvantageous for sustaining intermittent water loss.

Against the backdrop of such survival challenges, Mauricio Canals, Escuela de Salud Publica, Facultad de Medicina, Universidad de Chile, Chile, along with two research associates, carried out review research on the body design (morphological and anatomical) of many spiders found in Chile related to important features such as energetics, thermal selection, and water balance to correlate those physical and physiological traits to their ecological and behavioral characteristics. Arachnologists revealed that the preferred and critical temperatures of spiders vary with the variety and infraspecific levels of Chilean spider and are assumed to adjusted to the inherent phenotypic plasticity of the concerned species (Canals et al., 2015). An intriguing fact revealed by Fusco and Minelli (2010) is that undulations of environmental factors promote the development and evolution of metazoan life on Earth. They further advocated that the evolution of switches of alternative phenotypes was triggered by environmental cues and the temporal evolution of certain phenotypes among a batch of alternative phenotypes, which redefined the combination of genotype and newly preferred phenotype and attributed a unique twist to the evolution of any organism at species and infraspecific levels. During the ecophysiological investigations on the Chilean mygalomorphs, arachnologists noticed that the preferred temperature of *Paraphysa parvula*, a species that has adapted to the high altitudinal zone, is identical to that of *Grammostola rosea*, the species adapted to the lower altitudinal zone (Canals et al., 2015). The careful investigation further revealed that *P. parvula* (the high altitudinal one) has phenotypic plasticity, whereas *G. rosea* (of the lower altitude) does not have this property. If any alteration of environmental factors happens in both altitudinal locations or any accidental makeover could push both species to attain a new altitudinal location, the chance of achieving sustainability of *P. parvula* is most likely to be respondent toward the course of evolution by possessing the unique trait of phenotypic plasticity. Devoid of the inherent evolutionary switch of phenotypic plasticity, however, the Chilean *G. rosea* spider in a low altitudinal zone must cope with any altitudinal alteration in the potential change of environmental factors and might be at the brink of extinction in the worst possible situation. The ecophysiological investigations of Canals and his associates (2015) on the Chilean spider species in association to its body size and architecture attempted to elucidate the impact of the undulations of the Mediterranean environmental factors pointed to review of the natural adaptation measure adopted by Chilean spiders to ensure their sustainability by optimizing and reoptimizing the use of energy and utilizing their inherent trait of phenotypic plasticity to interact and reinteract with changing environmental factors to ensure sustainability in the ever-changing terrestrial habitat by traversing down the trail of biological evolution.

The *Theory of Island Biogeography* was the maiden effort to predict natural patterns of formation and functional pattern of species without considering their ecological niche (MacArthur and Wilson, 1963, 2001). These ecologists referred to the niche theory to elucidate the importance of environmental conditions in the distributional formation and abundance of species in community assemblage. In the widely debated but theoretically justified (but yet to be accepted above and beyond the threshold of controversy from a practical viewpoint) "Neutral Theory of Biodiversity and Biogeography" Hubbell (1997, 2001) tried to promote the ideas that the niche had of limited scope to attribute the formation and dynamics of any biological communities. Earlier ecological investigations ascertained that the niche theory had clearly distinguished the differences between species and the environmental factors and defined the species distribution, composition, and diversity of communities (Hutchinson, 1957; Schoener, 1989; Tokeshi, 1999). From a theoretical perspective, the contemporary framework of the neutral theory of biodiversity and biogeography has potentially defined the functional aspects of "ecological drift" that could elucidate the occurrence of species abundance distributions, species—area relationships, and the appraisal of changes in beta diversity. Scientists observed that in the neutral theory at the time of analyzing and interpreting the composition and diversity of communities, the main flaws were with the differences among the species that had completely been ignored, yet the predictions mostly fit to define the assemblage of the natural community (Hubbel, 2005). In 2007, Wiebke Entling, at Zoological Institute, University of Bern, Switzerland, along with four research associates, investigated the importance of environmental ingredients, such as light and shade conditions, availability of moisture, and the dynamism of habitat niche (diverse niche position and diverse range of niche width), to define the evolution of the habitat niche of spiders in the lowland habitats in Central Europe (comprising Austria, Germany, Belgium, Luxembourg, the Netherlands, and Switzerland). Entling and his research colleagues (2007) analyzed (by means of correspondence analysis) around 244 published spider communities from 70 distinct habitats of Central Europe and determined community scores to establish correlations with habitat features. The team of research ecologists used the species' scores to determine the niche positions represented by the mean scores and the standard deviation of the scores that represented the niche width to estimate variation in the taxonomic hierarchy of the spider communities in Central Europe by testing niche conservatism. The results of those correspondence analyses were quite intriguing (Entling et al., 2007):

a. A "hump-shaped" graphical relationship was found between the "niche width" and the environmental gradients (moisture and shade).
b. The narrow niches of the spider species found at the end of the environmental gradients (such as the diminishing level of shade and moisture) indicated that adaptations to such extreme habitats render constraints in ecological flexibility.
c. Habitat niche conservatism was found to be much lower than phylogenetic conservatism in spider body size.
d. Beta diversity of spider communities in the open forest areas in Central Europe was substantially higher than for spider communities in closed or dense forest areas.

Based on the key observations of that correspondence analysis, Entling and his colleagues reached two important conclusions:

1. Environmental factors were the key impetus that shaped the beta diversity of spiders across open habitat areas on a spatial scale.
2. The habitat niche of any biological species was assumed to evolve much faster than the evolution of any new morphological or physiological traits.

Henceforth, the contemporary correspondence analysis on the niche properties of Central European spiders conducted by Wiebke Entling and colleagues (2007) further indicated that "evolution of habitat niche" played a key role in the evolution and diversification of spiders on Earth since its journey initiated around 400 million years ago. Yet a minority of field ecologists raised an issue that this experimental application of the neutral theory of biodiversity and biogeography has failed to address the fact that species differ in their niche, so an important query arises, and that is whether an understanding of major patterns in nature is possible without integrating the niche into the details (Entling et al., 2007).

The silk properties of the spiderweb have been extensively studied for some time, but nevertheless, little is known about the thermal insulation properties of natural materials such as cocoons and the transitional developmental state of arthropods that plays an important role in the adaptation of any species in its juvenile stage (Hieber, 1985; Zhang et al., 2013). Eliodoro Chiavazzo, from Multi-Scale Modeling Lab, Politecnico di Torino, Italy, along with six other research scientists, coordinated an investigation on thermomechanical properties and an investigation of insulation properties of the spider sacs or cocoons of the European cave spider. They found that "*Meta menardi*" (belong to Tetragnathidae) took refuge in the "twilight zone available in most of the European caves" to get a better understanding of the female behavioral choice regarding the site selection for laying eggs to enhance the chance of spider survival during the juvenile stage. Generally, it has been observed that the females of European cave spiders lay a "drop-shaped egg sac (cocoon)," and the biomechanical investigations of the natural ingredient of this cocoon and the silk fibers have recently been investigated (Lepore et al., 2012). Arachnologists and evolutionary biologists have considered the biobehavioral traits that prompt the selection of suitable areas for laying eggs by female spiders to maximize egg hatching and ensure the highest juvenile survival rate as integrated into Darwinian fitness. Though its importance has been assumed, the factors prompting such a decision have not yet been clearly understood. To obtain a clear view of the selective forces that trigger the mature females of European cave-dwelling spiders to choose the depositional area, Chiavazzo and the coresearchers (2015) engaged in-field observational data for laboratory experiments. From the preliminary level of observations, Chiavazzo and his colleagues (2015) found that the diameter of the cocoon varied from 4 to 6 cm and contained 200–300 eggs laid adjacent to the entrance of the cave at the end of summer (Hörweg et al., 2011). The scientists observed that the newly hatched spiderlings remained in the cocoon until completion of their first molt and survived inside the cocoon and escaped unfavorable environmental conditions by feeding on the yolk of the eggs (Tercafs, 1988; Smithers, 2005).

Arachnologists and ecologists have contemplated that cocoons are designed to protect eggs from exposure to exterior thermal variations (undulating oscillations of hot and cold thermal expositions), desiccations (the changes in moisture content due to seasonal weather changes), and direct exposure to predators and parasites that would likely devour or parasitize the eggs that would stall the process of hatching (McCook, 1890; Hieber, 1985). The scientists further observed that the potential role played by the cocoon was as a thermal insulator of the eggs to avert the exposition of thermal undulations and extend constant microclimatic conditions to the newly hatched spiderlings and support them to take refuge inside the cocoon until spring would promote their embryonic development into mature spiders (Kaston, 1981; Turnbull, 1973; Schaefer, 1976). During spring, the grown-up spiderlings, after leaving the cocoon, traversed to the cave entrance and tried to avail themselves of the exposure of light, and the wind took them from the cave to travel far and wide from the location of their birth to reach distant habitats and occupy diverse ecological niches via air-driven logistics, called dispersal of air-plankton or ballooning (Smithers, 2005; Carylsue, 2015). The careful field observations of Chiavazzo and his fellow scientists (2015) led them to determine that the optimal velocity of air current and distance from the cave entrance to the spot of laying eggs selected by the female spiders in a particular cave had been rationally judged by the female spiders to ensure the optimal dispersal during spring after the spiderlings had emerged from the cocoon. The newly acquired bits of knowledge on the thermomechanical properties of natural materials such as the silk threads of the cocoons provided new impressions about the adaptation of prehistoric invertebrate species such as spiders and shows how a species could deal with environmental exposure and changes that would satisfactorily sustain its developmental stage, and it could lead to a new perspective about the evolution of spiders on Earth. To determine the potential methods for safeguarding spider eggs and newly hatched spiderlings by the egg sac or cocoons, Chiavazzo and his colleagues (2015) adopted an innovative investigative model called "entropy generation minimization methods" or EGM, the idea of which was adapted from the realm of applied physics/engineering science, to analyze the thermodynamic explanation generated through a heat transfer and the irreversibility of fluid flow mechanics. In the field of applied physics, EGM ensures minimal destruction or use of energy fuel to accomplish a task. By transposing the same principle within the biological realm, it has been observed that fitness of biological function entails the generation of heat transfer, and the maximization of mass transport and fluid flows when the required energy (which has been acquired from food intake) is minimized (Chiavazzo et al., 2015). A real-life instance of such thermodynamic arguments of entropy in the biological world has been provided by Bejan (1996, 2001), who presented that "optimization of food intake" (i.e., useful energy or exergy) can be associated with the optimal size of organs in animals or with bird flight.

The experimental observations of Chiavazzo and fellow scientists in 2015 determined that the females of the European cave spider select their preferential locations based on two environmental variables: (1) air velocity and (2) the distance from the cave entrance in a particular cave to the place selected for the deposition of eggs. The scientists further revealed how preferred sites are optimized:

a. Air velocity in the range of 0.3—0.6 m/s in the vicinity of the deposition site of the egg.
b. A range of 8—10 m in linear distance from the cage entrance.

The scientists determined that the environmental variables rendered such microclimatic conditions in the caves were helpful in the development of eggs of the European cave spiders inside the cave and optimized the migration of spiderlings outside the cave. Against the backdrop of the thermodynamic optimization argument, Chiavazzo and colleagues (2015) contemplated that the optimal values of those two environmental variables were correlated with the minimal level of irreversibility, and three key irreversibility sources were identified:

1. Fluid dynamics friction
2. Mass transport, and
3. Heat transfer

From the biological perspective of entropic argument, an increase in fluid dynamics irreversibility played an unwanted role—for example, the enhanced level of metabolic energy expenditure at the time the spiderlings left the caves in the course of bubble action driven by air current—as it has been observed that the larger the trail of migration the spiderlings traversed, the higher the volume of metabolic energy consumption that would be required. In a separate experiment, arachnologists observed that in the embryonic developmental stage, the yolks are consumed by the spiderlings inside the cocoon, and they stop feeding before taking part in the air-driven migration to leave the cocoon and the cave to travel to new ecological niches and resume feeding during colonization of new habitats (Smithers, 2005). Scientists have also observed that spiderlings that emerge from the cocoons are enticed by phototaxis movement to move from the dark environment of the cave to the cave entrance flooded with light (Tercafs, 1988; Smithers, 2005; Carylsue, 2015). According to the EGM argument, the absence of food intake in spiderlings, prior to emergence to the cave entrance shows that cocoons at the proximity of the cave entrance tend to minimize the consumption of energy in order to streamline migration from the cave and should be considered the primary effort of the spider to maximize its reproductive goal (Chivazzo et al., 2015). The next issue in this thermodynamic model is consideration of the microclimatic conditions at the surface and inside spider cocoons. According to the principles of fluid dynamics irreversibility, the friction between the surface of the cocoon and the blowing air around its surface should be minimized to avoid heat generation inside the cocoon that could potentially hinder the growth and development of the eggs inside. The experiment on cocoon thermal properties and EGM arguments define the choices of female members of the European cave spiders as being based on ecological conditions to choose cocoon deposition areas that maximize reproductive success, and from an evolutionary perspective, progression in the shifting paradigm according to the principles of Darwinian fitness (Chiavazzo et al., 2015).

References

Bejan, A., 1996. Entropy Generation Minimization. CRC Press, Boca Raton, New York, p. 361.

Bejan, A., 2001. Entropy generation minimization: the method and its applications. J. Mech. Eng. 47 (8), 345–355.

Canals, M., Veloso, C., Moreno, L., Solis, R., 2015. Low metabolic rates in primitive hunters and weaver spiders. Physiol. Entomol. 40 (3), 232–238. https://doi.org/10.1111/phen.12108.

Carylsue, May 20, 2015. Spidery 'Plankton' pours down. Natl. Geogr. https://blog.education.nationalgeographic.org/2015/05/20/spidery-plankton-pours-down/.

Černecká, L., Michalko, R., Krištín, A., 2017. Abiotic factors and biotic interactions jointly drive spider assemblages in nest-boxes in mixed forests. J. Arachnol. 45 (2), 213–222. https://doi.org/10.1636/JoA-S-15-005.1.

Chai, Y., Wilgers, D., 2015. Effects of temperature and light levels on refuge use in the wolf spider Rabidosa punctulata. Trans. Kans. Acad. Sci. 118, 194–200. https://doi.org/10.1660/062.118.0302.

Chiavazzo, E., Isaia, M., Mammola, S., Lepore, E., Ventola, L., Asinari, P., Pugno, N., 2015. Cave spiders choose optimal environmental factors with respect to the generated entropy when laying their cocoon. Sci. Rep. 5, 7611. https://doi.org/10.1038/srep07611.

Dahl, M.T., Yoccuz, N.G., Aakra, K., Coulson, S.J., 2018. The Araneae of Svalbard: landscape scale spider community diversity in the High Arctic. Polar Biol. 41, 839–853.

Entling, W., Entling, M., Bacher, S., Brandl, R., Nentwig, W., 2007. Niche properties of Central European spiders: shading, moisture and the evolution of the habitat niche. Global Ecol. Biogeogr. 16, 440–448. https://doi.org/10.1111/j.1466-8238.2006.00305.x.

Ferro, V.G., Romanowski, H.P., 2012. Diversity and composition of tiger moths (Lepidoptera: arctiidae) in an area of Atlantic Forest in southern Brazil: is the fauna more diverse in the grassland or in the forest? Zoologia 29, 7–18. https://doi.org/10.1590/S1984-46702012000100002.

Foelex, R.F., 1996. Biology of Spiders. Oxford University Press, p. 432.

Fusco, G., Minelli, A., 2010. Phenotypic plasticity in development and evolution: facts and concepts. Phil. Trans. R. Soc. B 365, 547–556. https://royalsocietypublishing.org/author/Minelli%2C+Alessandro.

Hieber, C.S., 1985. The 'insulation' layer in the cocoons of *Argiope aurantia*(araneae : araneidae). J. Therm. Biol. 10 (3), 171–175.

Hörweg, C., Blick, T., Zaenker, S., 2011. The large cave spider Meta menardi (Araneae:Tetragnathidae), spider of the year 2012. Arachnol. Mittl. 42, 62–64.

Hubbell, S.P., 1997. A unified theory of biogeography and relative species abundance and its application to tropical rain forests and coral reefs. Coral Reefs 16 (Suppl. l), S9–S21.

Hubbell, S.P., 2001. The Unified Neutral Theory of Biodiversity and Biogeography. Princeton University Press, Princeton, NJ, p. 392.

Hubbell, S.P., 2005. Neutral theory in community ecology and the hypothesis of functional equivalence. Funct. Ecol. 19 (1), 166–172. https://doi.org/10.1111/j.0269-8463.2005.00965.x.

Hutchinson, G.E., 1957. Concluding remarks. Cold Spring Harbor. Symp. Quant. Biol. 22, 415–427. https://doi.org/10.1101/SQB.1957.022.01.039.

Kaston, B.J., 1981. Spiders of Connecticut. Department of Environmental Protection, State Geological and Natural History Survey of Connecticut, p. 1020.

Lepore, E., Marchioro, A., Isaia, M., Buehler, M.J., Pugno, N.M., 2012. Evidence of the most stretchable egg sac silk stalk of the European spider of the year Meta menardi. PLoS One 7, e30500.

MacArthur, R.H., Wilson, E.O., 1963. An equilibrium theory of insular zoogeography. Evolution 17 (4), 373–387. https://doi.org/10.1111/j.1558-5646.1963.tb03295.x.

MacArthur, R.H., Wilson, E.O., 2001. The Theory of Island Biogeography. Princeton University Press, Princeton, NJ, p. 203.

McCook, H.C., 1890. American Spiders and Their Spinning Work, vol. II. Allen, Lane, and Scott, Philadelphia, p. 480.

Melo-Braga, M.N., De Marco Almeida, F., Dos Santos, D.M., de Avelar Júnior, J.T., Dos Reis, P.V.M., de Lima, M.E., 2020. Antimicrobial peptides from Lycosidae (Sundevall, 1833) spiders. Curr. Protein Pept. Sci. 21 (5), 527–541. https://doi.org/10.2174/1389203721666200116091911.

Primer, A., 2004. Phenotypic variation from single genotypes. In: DeWitt, T.J., Scheiner, S.M. (Eds.), Phenotypic Plasticity: Functional and Conceptual Approaches. Oxford University Press, Oxford.

Rodriguez-Artigas, S., Ballester, R., Corronca, J., 2016. Factors that influence the beta-diversity of spider communities in northwestern Argentinean Grasslands. PeerJ 4, e1946. https://doi.org/10.7717/peerj.1946.

Schaefer, C.W., 1976. The reality of the higher taxonomic categories. J. Zool. Syst. Evol. Res. 14 (1), 1–10. https://doi.org/10.1111/j.1439-0469.1976.tb00513.x.

Schoener, T.W., 1989. The ecological niche. In: Cherrett, J.M. (Ed.), Ecological Concepts: The Contribution of Ecology to an Understanding of the Natural World, Symposium British Ecological Society. Blackwell Scientific Publications, Cambridge.

Simó, M., Laborda, A., Jorge, C., Guerrero, J.C., Dias, M.A., Castro, M., 2011. Introduction, distribution and habitats of the invasive spider *Badumna longinqua* (L. Koch, 1867) (Araneae: desidae) in Uruguay, with notes on its world dispersion. J. Nat. Hist. 45 (27–28), 1637–1648. https://doi.org/10.1080/00222933.2011.559599.

Smithers, P., 2005. The early life history and dispersal of the cave spider *Meta menardi* (Latreille, 1804) (Araneae: Tetragnathidae). Bull. Br. Arachnol. Soc. 13, 213–216.

Soininen, J., McDonald, R., Hillebrand, H., 2007. The distance decay of similarity in ecological communities. Ecography 30 (1), 3–12. https://doi.org/10.1111/j.0906-7590.2007.04817.x.

Stańska, M., Stański, T., Wielgosz, E., Hajdamowicz, I., 2018. Impact of habitat complexity on body size of two spider species, *Alopecosa cuneata* and *A. pulverulenta* (araneae, Lycosidae), in River Valley grasslands. Pol. J. Environ. Stud. 27, 853–859.

Sunderland, K., Samu, F., 2000. Effects of agricultural diversification on the abundance, distribution, and pest control potential of spiders: a review. Entomol. Exp. Appl. 95, 1–13. https://doi.org/10.1023/A:1003986225443.

Tercafs, R., 1988. Nesting dispersal in a cave spider. Mem. Biospeleol. 15, 45–48.

Tokeshi, M., 1999. Species Coexistence: Ecological and Evolutionary Perspective. Blackwell Scientific, p. 454.

Turnbull, A.L., 1973. Ecology of the true spiders (Araneomorphae). A. Rev. Ent. 18, 305–348.

Viera, C., 2011. Comportamiento constructor y depredador de arañas de tela. In: Viera, C. (Ed.), Arácnidos de Uruguay. Diversidad, comportamiento y ecología. Montevideo: Ediciones de la Banda Oriental.

Whittaker, R.H., 1960. Vegetation of the siskiyou mountains, Oregon and California. Ecol. Monogr. 30 (3), 279–338. https://doi.org/10.2307/1943563.JSTOR.

Wilczek, G., Karcz, J., Putko, A., Kedziorski, A., Wilczek, P., Stalmach, M., Szulińska, E., 2017. The effect of ingested cadmium on the calorific value and structural properties of hunting webs produced by *Steatoda grossa* (Theridiidae) spiders. Sci. Total Environ. 586. https://doi.org/10.1016/j.scitotenv.2017.02.143.

Wise, D.H., 2006. Cannibalism, food limitation, intraspecific competition, and the regulation of spider populations. Annu. Rev. Entomol. 51, 441–465. https://doi.org/10.1146/annurev.ento.51.110104.150947.

Zhang, Z.-S., Li, S., Pham, D.-S., 2013. First description of comb-tailed spiders (araneae: hahniidae) from Vietnam. Zootaxa 3613 (4), 343–356.

CHAPTER TWELVE

Shifting of environmental factors related to ecophysiological and adaptive responses of spiders

Abstract

Scientists have witnessed more than one phenotypic expression from certain organisms with distinct genotypic profiles synced to respond to environmental undulations within their surroundings, a trait recognized as phenotypic plasticity. Scientists consider arachnids predominant invertebrates, organisms that evolved around 400 million years ago. In the evolutionary history of life, the role of phenotypic plasticity has recently been studied, and scientists have argued that the nonplastic life history of prehistoric and extinct organisms points further to the mismatch of the adaptive potential of old organisms and that environmental undulations likely expedited the extinction of several animals or created numerous living fossils. Prehistoric organisms such as spiders, predominantly attributed with phenotypic plasticity, have wide-open chances to traverse the trail of biological evolution. They are biologically competent enough to survive for long periods to become accustomed to microevolutionary changes.

Keywords: Arachnids; Environmental undulations; Microevolution; Phenotypic plasticity.

Both aerobic and anaerobic metabolism were the lowest, but the level of ATP was kept high in spiders from areas where heavy metals dominated as the stressing factors. In Jaworzno [Southern Poland], where organic and gaseous pollutants predominate, all Araneidae species showed higher aerobic activity, while Linyphiidae rose anaerobic activity, but maintained high levels of ADP and total adenylate pool with the lowest AEC index (below 0.7). Metal burdens were species-dependent and correlate well with concentrations of metals in spiders' milieu. The participation of other interacting pollutants in overall effect on biochemical and physiological changes in spiders depend on their total concentration, chemical form, and interdependencies on various organization level of an organism of spider.

G Marczyk, P. Migula & E. Trzcionka

Spider Evolution
ISBN: 978-0-323-90041-6
https://doi.org/10.1016/B978-0-323-90041-6.00009-7

Spiders are ubiquitous invertebrates occupying nearly all possible terrestrial ecosystems within every continental biome except Antarctica. With a diverse range of habitats, spiders inhabit terrestrial habitats, caves, trees, and arboreal ecosystems. Under inherent traits like phenotypic plasticity, the spider has a wide range of adaptations, so it has shown a penchant for spatiotemporal exploration of new climates and habitats since its origin and evolution on the Earth—which is estimated to have happened around 400 million years ago (mya). Against the backdrop of environmental undulations, an inherent genotypic profile of any organism producing more than one phenotypic expression is called phenotypic plasticity (West-Eberhard, 1989). It should be apparent that this phenotypic plasticity is distinct from phenotypic switching, where the occurrence of phenotypic changes happens without inducement by any environmental cues (Fusco and Minelli, 2010). Evolutionary biologists have observed that the extent of the plasticity of any organism is age-dependent, and they have further observed that when a suitable occurrence of environmental undulations is synced with the dynamic life history of the concerned organism, the occurrence of phenotypic plasticity is most likely to occur and displays its expression against the stern expression of the potential of the genotype (Pigliucci, 2001; Gabriel, 2005; Auld et al., 2010; Panchanathan and Frankenhuis, 2016).

The most intriguing part of the functional mode of expression is that the expression of any time-bound occurrence of phenotypic plasticity could be extended to the attribution to an incident of lifelong plasticity to the organism if its functionality is integrated via morphological or behavioral changes of the concerned organism (Wikelski and Thom, 2000; Forsman, 2015). Experimental models have helped scientists determine that long-lived organisms sustained under undulating environmental conditions are found to have more plasticity, and empirical investigations have ascertained that slow life histories or life spans correlate with larger brain sizes or higher learning potential (mammals, birds, butterflies, etc.) (Kokko et al., 2001; Gonzalez-Lagos et al., 2010; Shulz and Dunbar, 2010). But the direction and multitude of aging and occurrence of phenotypic plasticity have yet to be resolved in the arena of evolutionary biology as scientists could not make any unanimous decision whether a long life span would create any analogical penultimate jumping board that rendered any organism to be benefitted out of lifelong plasticity or the occurrence of reversible plasticity would enhance the extent of longevity (Borges, 2009). Through a series of hypothetical and theoretical perceptions has promulgated that life-span is the progenitor of patterns that eventually evolves plasticity (Botero et al., 2015; Tufto, 2015), yet the role of plasticity in the evolutionary history of life is understudied, and the nonplastic life history of the prehistoric and extinct organisms indicated that the adaptive miss-match of the old-organisms and the environmental undulations might expedite extinction of several animals or create numerous living fossils over time (Cotto and Ronce, 2014). However, the perceptional causality of the reversible phenotypic plasticity of any organism expresses its potential to go through the phenotypic changes that happen throughout its life and can be selected in the dynamic environment that also goes through changes in parallel to the passage of the life span of the concerned organism (Ratikainen and Kokko, 2019). Scientists have contemplated that if the extent of mismatch between the environmental changes and the nonplastic traits of an organism remains at a standstill or might be enhanced over time, it would enhance the rate of senescence. To understand the functional causality of the reversible plasticity, which usually covary with life span, the causality of the evolution of plasticity has been analyzed and contemplated in a unidirectional way triggered by the favorable undulations of environmental factors regarding the life history of the concerned organism. It seems it has a perceptional analogy of directional causality between eggs and chicken, whether chicken from eggs or eggs from chicken. Digging deeper, scientists have expressed their view in the current perspective that if the life span evolves caused by the plasticity trait, then the long span of life must not be integrated into the life-long plasticity. Thoughtful analysis and careful review of Ratikinen and Kokko's (2019) work figured out that causality should better be recognized as bidirectional as the plasticity, select for the longevity of life span. Furthermore, Ratikainen and Kokko (2019) said, *"Highly autocorrelated environmental fluctuations predict low investment in reversible plasticity and a phenotype that is reversible plasticity and a phenotype that is poorly matched to the environment at older ages. Such environments select for high reproductive effort and short life spans."*

However, in 2013, Ana Llandres from Institut de Recherche sur la Biologie del l'Insecte, Universities de Tours, France, along with four research associates, has worked on the impact of habitat heterogeneity on spiders that occurs within an individual's lifetime that supposedly favor the evolution of reversible plasticity of it (Llandres et al., 2013). Field ecologists and animal biologists have observed that environmental and hormonal factors play an essential role in the reversibility of colors in many animals, and the change in body colors helps animals to survive through many functional ranges from adaptive functions like crypsis for camouflaging in a particular background of any microenvironmental habitat adapted for survival or hunting or social interactions and thermoregulation, which helps to control its body temperature and is optimal for performing necessary physical and physiological functions. Though Llandres et al. (2013) tried their level best to investigate phenomena like the occurrence of reversible color changes in lower groups of animals such as spiders to comprehend their physiological and evolutionary impact on the concerned organisms, they agreed that substantial progression had occurred, but their findings have yet to define the underlying mechanisms that must occur at the molecular level in the future. To discover the effect of external environmental cues and internal hormonal undulations on the morphological change of colors of spiders, Llandres et al. (2013) have adopted an experimental model with the model organism Crab spider, *Thomisus onustus*, that has a preexisting record of color changes in different occasions and the metabolites produced during such changes, in the diverse Spatio-temporal background. Llandres et al. (2013) reviewed such dynamics of color changes using a spectrophotometer for a specific time (15 days) to put spiders in yellow-and-white-colored containers under natural light conditions. The chemical metabolites of spiders have been monitored with HPLC or High- Performance Liquid Chromatography. In their critical observations, Llandres et al. (2013) have monitored the hormonal control of color changes by injecting 20-hydroxyecdysone (20E) into the bloodstreams of spiders. The scientists found that spiders placed in yellow containers changed their color in yellow as it has stated synthesizing yellow pigment, whereas spiders placed in white container changed to white color and it has further been observed that a monochrome precursor, 3-OH-kynurenine, is found to be the key pigment for yellow coloration. The examination of Llandres et al. (2013) further revealed that the effect of 20E was identical to that when spiders were placed in yellow containers under natural light conditions. In most cases, however, changing the reversible morphological colors of animals relative to background objects or environment helped scientists define such cryptic changes in color as an effort to avert the risk of potential predators such as fish, chameleons, and cephalopods (Stevens and Merilaita, 2008). By and large, the exact reason behind reversible color charges of crab spiders remained hypothetical, as crab spiders are hardly sustained, and predatory attacks from the vertebrate predators in the wild habitat remained restricted, though it has often been attacked by spider wasps and mud-daubers (Morse, 2007). Many scientists have hypothesized that the reversible changes of color witnessed in crab spiders likely protect them from the potential risk of radiation (Llandres et al., 2013). It is quite intriguing that many evolutionary biologists have further postulated that crab spiders have evolved with the trait of changing cryptic coloration to match their background to that of their prey (Gabritschevsky, 2005; Rabaud, 1919).

Llandres et al. (2013) finally concluded that the "hormonal manipulations" are responsible for reversible changes of color of the Crab spiders. So, indicating the evolutionary importance of the adaptive changes, Llandres further said: *"It is nevertheless important to highlight the plasticity might be favored globally, even when fitness is not highest at any particular site."* (Llandres et al., 2013).

Evolutionary biologists have witnessed that terrestrialization of prehistoric arthropods have been taken place in Upper Cambrian- Early Ordovician, around 500 mya, though the erstwhile paleontological evidence indicated its earliest instances of origin (where it is terrestrial in origin is not very clear) in the middle Ordovician; but the clear morphological evidence of terrestrial arthropods are there during Silurian, which is around 400 mya (MacNaughton et al., 2002). So, the evolutionary biologists have contemplated that arthropods have gone through the adaptation, down the evolutionary labyrinth of terrestrialization in this big gap of 100 years (from 500 mya to 400 mya) to come out of aquatic habitat to settle down on the terrestrial habitat (Canals et al., 2015). To get through the bottleneck of transmigration from water habitat to terrestrial habitat, most of the animals have gone through three main key challenging issues:

a. The animals living in the terrestrial habitat need to adjust to huge fluctuations in temperature, humidity, available drinking water and oxygen, wind velocity, thermodynamic alterations, etc.

b. The concentration of water inside the body fluid of the transmigrating animals is found to be higher than outside of their environment, so they need to overcome the potential challenge of water loss and

c. In the prehistoric transmigrating animals, it has generally been observed that their smaller body size varies from a few millimeters to a few centimeters, and their body mass ranges from a few milligrams to a few grams, resulting from a high index of surface/volume ratio that renders them sustaining an extreme level of environmental swings in terms of variation of temperatures and humidity (Nentwig, 2013).

It has generally been observed that arthropods are smaller in size, with wide variations in their sizes. Paleontological investigations on the fossilized arthropod Gigantostracos by arachnologists have revealed that the body size of the arthropods reached around 3 cm during the Silurian period, and the gigantism mode of diversification of the arthropods has been observed in parallel to the high level of oxygen concentration existing in the congenial paleoenvironment at that time (Canals et al., 2015). The evolutionary biologists observed that the evolutionary surge of giant forms of prehistoric insects evolved in the hyperoxic environment at least twice, for the first time during the Carboniferous–Permian regime, followed by the middle of Jurassic–Cretaceous–Tertiary regime (Dudley 1998; Falkowski et al., 2005).

Nonetheless, the primary reason is that after traversing a long path of evolution and diversification, spiders appeared to be an omnipresent representative of invertebrates traced out in the microclimatic refugia in the lava tubes of Hawaii Islands (e.g., *Lycosa howarthi*) or in the scorching hot ambiance of tropical ecosystems of the sub-Saharan desert (e.g., *Cebrennus rechenbergi*) to the ice-cold caverns of temperate and arctic Tundra of Yellowknife in Canada (e.g., *Alopecosa aculeata*) (Hadley et al., 1981; Foelix, 2011). The key objective in ecophysiology is to define how a group of animals could sustain extreme environmental conditions such as high altitude cold arctic mountains or extremely hot desert ecosystems and how the physiological constraints impacted the biological adaptations (Bozinovic and Canals, 2008). Scientists observed that the structural features, behavioral, physiological, and biochemical characteristics of any organism, foraging, digestion, the thermodynamic efficiency of the concerned organism, and its pattern of energy production are found to be under selective pressure, which helped it to achieve sustainability in a place for a particular time (Canals et al., 2015). To support it scientifically, scientists have referred to an instance of a mammal species acclimatized to survive in extremely cold conditions mainly depends on its thermogenic ability, whereas the production of some litter and its quality (in terms of health and immunological standard) depends on the potential of its rate of energy production in milk yield by the female (Veloso and Bozinovic, 1993; Veloso, 2003). Based on such observations, evolutionary biologists have postulated that natural selection would promote such life strategies to ensure that equilibrium persists between the energy costs and energy inputs, assimilated from food, taken by any organism (Bozinovic et al., 1985). So, in terms of the accounting of energy dynamics, it is convincible that the organism which has the potential in maximizing the difference between "energy input" and "operating costs" would be favored by the natural selection, as the energy losses are found to be disadvantageous to achieve sustainability and reproductions (Canals et al., 2015). Technically, the adaptive process acts as a buffering mechanism to optimize the energy logistics and maintains the equilibrium status of the energy dynamics between energy inputs and energy outputs, and it creates an ambiance for the organism that would ensure the survival of the concerned organism to fit in the changing environment.

In terms of thermodynamics, spiders are considered ectothermic organisms whose internal temperature is controlled by the heat transferred from their external ambiance, and it has also been considered poikilotherms, the variation of the body temperature has been regulated by its behavioral changes (Canals et al., 2015). Scientists have observed that the variation in the temperature of poikilotherms is integrated to change its rate of metabolism, and the rate of metabolism is enhanced with the exponential enhancement of the temperature. In endothermic animals, scientists noticed that the maintenance of body temperature has always been maintained above the environmental temperature, so the heat loss from the outer body surface of any endotherms can be minimized. Hence, the body surface of animals varies in its outer appearance and is usually modified by several factors like food intake, locomotion, respiratory gaseous exchange, and heat exchange that led animals to adopt various geometric forms—Euclidian, cylindrical, spherical, etc. (Canals et al., 2015). By employing the behavioral changes, animals could change their geometry, like the tendency to form spherical geometric shapes behaviorally adopted by birds and mammals when exposed to cold in winter or sudden exposure to cold waves (Canals et al., 1989, 2011). Though terrestrial arthropods like spiders cannot change their outer surface because of their exoskeletons, they change their body positions to control the heat exchange process. Very recently, Grossi and Canals (2015) have tried to correlate the energy consumption and the occurrence of sexual size dimorphism of spiders, and they hypothesized that the cost of transport, as well as energy expenditure in locomotion and speed of the male spider, prompted them to promulgate that these traits are favored under selection pressure to attribute cutting the energy expenditure in the travel costs, incurred by the smaller sized (in comparison to their female counterparts) male spiders. From the perspective of ecophysiological adaptations, spiders with smaller body sizes enjoy several benefits that help them survive in the environment by increasing their surface for oxygen exchange and an enhanced level of body surface area per volume unit, offering them better wind resistance, and favoring their dispersal by passive flight, but their small sizes cause disadvantages such as fast temperature loss a higher rate of water loss, which are counterbalanced by the metabolic and biobehavioral adaptations of spiders (Canals et al., 2015).

Many arachnologists have observed that the low level of resting metabolic rate can be assumed to be associated with the rate of adaptation to environments with low predictability and low levels of prey availability (Greenstone and Bennett, 1980; Wilder, 2011). Arachnologists have tried to present a physiological explanation of such adaptive features of spiders as they mentioned that spiders used hydrostatic pressure to extend their appendages rather than active muscles to a lesser extent with consequent metabolic activity (Anderson and Prestwich, 1982). Arachnologists have tried to explain further that the low level of resting metabolic rate indicated spiders' unique feature that helped them survive even without food (Tanaka and Ito, 1982; Nentwig, 2013). Many animal biologists have construed that spiders have reduced the rate of metabolism in parallel to the scarcity of food level (Ito, 1964; Tanaka and Ito, 1982; Canals et al., 2011). In the studies of spider biology, arachnologists did not have enough hands-on research reference on the thermal biology of this tiny creature, as apparently, it seems that there is a close relationship between the cardinal limit of thermal tolerance of the spider and its habitat (Hanna and Cobb, 2007). Ecologists and invertebrate biologists have observed that environmental temperatures play an important role in the lives of spiders in the concerned microenvironment. As scientists noticed that thermal tolerance limit not only provoked spiders in the exterior environment to make them hyperactive for foraging, it also intrigued them to look for an appropriate mating partner in the mating season, mainly to look for the females, take refuge in the shelter and it also triggered the females to take rest in the shelter with their egg sac (Hanna and Cobb, 2007). In a unique experiment, De Vito and Formanowics (2003) observed that the heat shock endurance level of *Pirata sedentarius* in the juvenile stage is substantially higher than in their adult stage. Many thermal, biological investigations have been conducted on the females of *Paraphysa parvula*, a tiny mygalomorph spider inhabiting the Andes mountain ranges, and many scientists noticed that the females with egg sacs had 3°C lower than their nonbreeding counterparts, though another group of scientists hypothesized that the preferred temperatures of the reproductive females are considerably higher than the nonreproductive females (Vlijm et al., 1963; Frick et al., 2007). A recent thermoregulatory investigation on *P. parvula* has revealed that it changes its body temperature to maintain a similar temperature around its substrate, air, or stone of the inhabiting cave, and a similar biobehavioral act of thermoregulation has been observed in another spider species, *Aphonopelma* sp. Scientists have generally observed that as the temperature of any microenvironment increases, that rendered the inhabitant spider to sustain evaporative water loss. A recent investigation on the *P. parvula* shows that at 40°C, the evaporative water loss has increased 10 times, so it should be considered thermal limit and scientists have also estimated the evaporative water loss during the act of respiration through book lung and which has been estimated around 60% (Figueroa et al., 2010).

Ecologists and animal biologists have observed that the endurance potential of desiccation tolerance has a strong correlation with the habitat specificity and the physiological and behavioral adaptation of spiders (Vollmer and MacMahon, 1974; De Vito et al., 2004). To determine a comparative account of the endurance level of the heat stress and desiccation stress on the three species of wolf spiders (Lycosidae) *Pirata sedentarius*, *Pardosa lapidicina*, and *Pardosa fuscula* and tried to correlate it with their microhabitats, De Vito et al. (2004) have conducted an investigation and scientists noticed that *P. sedentarius* has a remarkable low tolerance level of endurance, that rendered it ecologically endemic to the microhabitat presented at the proximity of water bodies, in comparison to the two other Wolf spider species that are well adapted in heterogeneous habitats and pan-endemic in distribution. The animal biologists observed that the breathing gas exchanges in the environment happen in spiders, specifically the mygalomorph spiders through the book lungs (where a hypodermic gas barrier separates atrium from hemolymph), the key respiratory organ, along with tiny capillary, cuticular, porous surface exchange respiratory systems called spiracles (Canals et al., 2008, 2015); opening and closing of the spiracles is a complex physiological process. Opening of the spiracles has more frequently been observed at high temperatures at increased physical actions when the concerned organism needs a high oxygen demand level (Schmidt-Nielsen, 1997). The most intriguing fact that scientists have witnessed is that the concentration of carbon dioxide level in the air controls the ventilation of the respiratory system of spiders, specifically controlling the opening of the spiracles, through which the oxygen is absorbed by the body surface of spiders (Schmidt-Nielsen, 1997). Though it has not been examined so far on spiders, the variation of carbon dioxide concentration on the spiracular respiration of arthropods, like the cockroach, revealed that at around 1% concentration of CO_2 spiracular respiration get started as spiracles start opening, whereas at 2% the spiracles are relatively open and at 3%, the spiracles remained wide open (Schmidt-Nielsen, 1997). However, spiracular respiration is important in the respiratory action of insects and spiders in terms of achieving sustainability when the organism has to cope in an undulating environmental condition to be alive. But it has also been observed that wide-open spiracles would render the intermittent evaporative water-loss of the concerned organism for a long time, which might be physiologically most disadvantageous action, causing irreversible dehydration and death of the concerned organism (e.g., earthworm) (Schmidt-Nielsen, 1997). So, the adaptive molecular switch which regulates the turn-on/off mechanism of spiracular respiration ensures the survival of these tiny invertebrate arthropods like insects, spiders, etc., by acting as a fitness pressure to promote evolutionary progress and diversification of spiders over time.

Social behaviors among arachnid members are diverse, as some spiders form a colony and prefer to live like social spiders, whereas others prefer to lead solitary lives except for intraspecific interactions like fighting and mating. Irrespective of its way of living, the main ecological equilibrium that maintains the population diversity of spiders and other insects is the predator–prey relationship. We observe that the primary niche of spiders in every ecosystem is universal, that is, controlling the insect population by eating them. According to the conventional model of energy trafficking through trophic, spiders are a good source of food to a wide array of large-sized predators ranging from bigger-sized wasps, carnivorous birds, lizards, and even small mammals. On the other hand, spiders are found to be the key predators of a wide array of insects either directly engaged in the hunting of prey (e.g., wolf spiders) or coordinated information of webs to trap the insects in its web to kill them and eat them (e.g., orb-weavers). From the perspective of biological control of insects, spiders are considered to be the most sustainable option to the agriculturists and horticulturists as it is the feasible, environment-friendly alternative for controlling the pests in the cultivated areas, instead of using chemical insecticides and pesticides (which are toxic to biological organisms and affect the normal living of organisms in the concerned ecosystems), intermittent use of which stalled the biogeochemical cycles of any environment and that not only jeopardize the equilibrium of biosphere but enhance the rate of extinction of certain species, of the concerned ecosystem. Scientists of the Colorado University observed that the control of the population of spiders is more or less biologically self-regulated, as the surging population of any spider is often killed and devoured by other groups and often by the same species (Demarino, 2017). In the year of 2003, a group of arachnologists from the University of Michigan observed that introducing or exterminating any species of spider did not play an important role in controlling the insect population of any ecosystem but the existence and interaction of many spiders found to be effective in controlling the spread of pests in an ecosystem (Demarino, 2017). Hence, the existence of the heterogeneous community of spiders, rather than a selective or homogenous community of spiders, should be regarded as the ecosystem stabilizer of the concerned ecosystem, particularly the ecosystems (e.g., land of cultivated crops, fruit orchards, horticultural fields, etc.), and the abode of a huge number of pests. Regarding the evaluation of prey capture potential of the colonial spiders (the three spiders of *Metepeira* sp. from Mexico), a unique investigation has been done by George Uetz, Department of Biological Science, University of Cincinnati, USA, and he observed the enhancement of the prey capture potential of it as an effect of "ricochet effect"[1] (Uetz, 1989). In his concurrent investigation, Uetz (1989) witnessed that different species of *Metepeira* occupied different habitats, and each species had a diverse level of social organization (number of individuals vary in the distinct form of community). Uetz (1989) observed that the newly discovered species of the Colonial spider, *Metepeira atascadero*, usually found in solitary or small groups living in grassland habitat, has gained somewhat from the impact of prey ricochet effects, and Uetz (1989) has further noticed that the rate of prey capture by this species is very low, but the variance in prey capture by each spider is remarkably high. In his observation on *M. spinipes*, which has inhabited the mesic sites of agricultural plantations, groups of 10–150 individuals in communities have been attributed to the higher number of capture prey with larger prey sizes by virtue of the "ricochet effect." In the third species of Colonial spider, *M. incrassata*, Uetz (1989) has witnessed that this species has formed the community, comprised a greater number of individuals (100–1000), and occupied the tropical rain forests and agricultural plantations and the prey capture potential of *M. incrassata* has similarly been influenced by the "ricochet effect." It is quite interesting to notice that irrespective of the ricochet effect, the taxonomic composition of the prey in *M. incrassata*, the species with the highest number of individuals at the community level, shows an affinity toward the insects belonging to Lepidoptera that has not been observed in *M. atascadero* and *M. spinipes*. It seems that the species diversity of Lepidoptera is high in the tropical ecosystem, hence by the ricochet effect, the capture of this prey is very high in *M. incrassata*. A comparative account of prey capture between *M. atascadero* (Colonial spider with the lowest impact of ricochet effect) and *M. incrassata* (Colonial spider with the highest impact of ricochet effect) revealed that the ricochet effect enhanced the prey capturing efficiency of the predators across the wide array of prey insects and this kind of acquired behavioral functions would found to be fruitful in the

[1] Ricochet effect: Ricochet is an act of rebound, bouncing (an effect of skip off the surface) that typically occurs with a moving particle in the case of a projectile. Normally it happens with shooting bullets and may cause collateral damage to any object in its surrounding due to ricocheting bullets and bullet fragments.

enhancement of the fitness of any species down the trail of its evolutionary journey (Uetz, 1989).

Most of the time, it has been noticed that the biological organism, particularly the smaller ones like spiders, would go through the rapid adjustment to cope with environmental changes, even in a much better way than the flowering plants or lichens do. In this concern, the concept of biotope needs to be elucidated further to understand the ecological functions and evolutionary progress of any biological organism in its Spatio-temporal background. A biotope is defined as a small geographical place, ecological abode, or habitat that supports its distinctive community of plant and animal species, e.g., the bark of a Sugar Maple tree acts as a microhabitat and supports a specific type of lichens, lepidopteran insects, and spiders in the Boreal Forest system of North America. However, the term "Biotope" has been used by the great ecologist Friedrich Dahl in 1908 for the very first time (De Bont and Lachmund, 2017). From the conventional practices of analysis of the flora, vascular plants of any specific habitat have been considered the general ecological state of the biotope. As ecologists have noticed, distinct types of biotopes are integrated into certain spider faunas, as the distinct type of vascular plants is a specific part of the concerned biotope. Hence, ecologists have chosen to get themselves engaged in analyzing the spider fauna rather than analyzing the vegetation in which that distinct type of spider fauna integrated with any specific habitat in their ecological analysis of biotopes to get a better picture of the spatial structure and microclimate of the environment concerning its integration to its distinctive plant and animal communities in it and understanding the dynamics of biological communities of that specific habitat (Barnes and Barnes, 1955; Curtis, 1978; Schaefer, 1978; Heublin, 1982). But it does not necessarily mean that there is a strong correlation between the number of spider species and the number of vascular plants in any microclimatic condition of any habitat in an environment, rather synecological investigations ascertained it as an insignificant factor, but there is a strong positive correlation noticed between the number of spider species and the percentage cover of the vascular plants (Duffey, 1968; Robinson, 1981; Borgesen et al., 1984). Hence, arachnologists have considered that considering spider fauna is a plausible alternative to consider instead of considering the vascular plants during the characterization of biotopes.

For a long time, since the Industrial Revolution in Britain and even before that, lichens have been considered the ecological indicator of pollution indicator to help scientists appraise the level of air pollution of any particular geographical location at the spatial level. Though such preexisting trend of investigations is not there, yet the zoologists and environmental scientists engaged in studying the impact of SO_2 pollution of the spider faunas on trees, and in the preliminary level of investigations, it has generally been observed that the density of spiders is negatively correlated to the SO_2 level in the air in its ambiance though many scientists have contemplated that correlation is not significant enough (Gilbert, 1971; Andre 1977; Clausen, 1984). However, a remarkable change in the population density of spiders is negatively correlated to the extent of pollution of the concerned habitat that the spider populations belonged to. The relationship of SO_2 concentration in air and the spider density population has considered being indirectly related as the decreasing of prey density is related to the lower level of primary productivity of trees that have been affected by the enhanced level of SO_2 concentration in the air (Varshney et al., 1979). So, the relative density of spiders increases (with respect to arthropod prey, the insects) if the number of insects goes down, and it shows that SO_2 concentration in air and the relative density of spiders are directly correlated, rather than establishing any direct relationship between the absolute density of spiders and the SO_2 concentration in air (Andre, 1977; Clausen, 1984).

Apart from air pollutants, several investigations have been conducted on the effect of metal pollution on the spider faunas, mainly inhabited in caves and mines. Clausen (1984) has revealed the positive correlation between the relative frequency of *Clubiona* sp. (in respect to other species of spiders) and the lead (Pb) burden in the atmosphere. To determine the lead burden and the other heavy metal pollutants in the atmosphere, scientists collected the lichens (e.g., *Lecanora*) from the zone of pollution and examined the concentration of heavy metal from it to determine the atmospheric heavy metal burden (Pilegard, 1978; Moseholm, 1981; Clausen, 1986). In the same way, the atmospheric lead burden has also been determined by measuring the Pb, Cu, Zn level in spiders (e.g., lead concentration has been measured in the spider *Araneus umbraticus* in northeastern New Zealand and Denmark) by Price et al. (1974), Wade et al. (1980), Bengtsson and Rundgren (1984). Bengtsson and Rundgren (1984) have observed similar observations on the spider fauna belonging to Clubionidae for the heavy metals of Cu (Copper), Pb (Lead), and Zn (Zinc). The presence of these heavy metals and chemicals in excessive levels in the environment in the ambiance of any biological organism like a spider would inevitably change its physical, physiological, metabolic, ecological, and genetic characteristics, its way of living, and its interactions in intra-and interspecific level, often put them in the brink of extinction or expedite its mode of evolution in nature.

In 1993, scientists investigated the effect of environmental pollutants on five spider species (with a diverse range of taxonomic identity, behavior, spatial distribution, hunting strategies, web construction patterns, etc.) at Jaworzno in Southern Poland, an area with a high extent of organic and gaseous pollutants. It has been observed that under the high exposition of pollutants, spiders belonging to Araneidae responded with a high rate of aerobic metabolism, whereas spiders belonging to Linyphiidae responded with a higher rate of anaerobic metabolism and maintaining a high level of ADP (adenosine diphosphate), adenylate pool with the lowest AEC index (>0.7) (Marczyk et al., 1993). It has further been observed that the metal burden potential of spiders living in the polluted environment varies from species to species. In a contemporary investigation, scientists investigated with a single or multiple doses of dimethoate (OP pesticide) on spiders like non-web-building wolf spiders (*Pardosa lugubris* of Lycosidae) and funnel-web (*Agelena labyrinthica* of Agelenidae) which have been collected from polluted meadows and made them exposed to diverse exposure of OP pesticides (Babczynska et al., 2006). In that study, an effort has been made to determine whether an intermittent exposure of heavy pollutants in the habitats under a strong industrial environment would rend it —hyper-activate or hyper-energized the biochemical defensive safeguards to survive in the environment with hyper-chemical stress. Scientists have used the detoxifying (like carboxylesterase, glutathione S-transferase) and antioxidative enzymes (like Selene dependent glutathione and Selene independent glutathione peroxidases) to observed that *P. lugubris* found to be more resistant to the higher level of pesticide intoxication, whereas no such result has experimentally been found on *A. labyrinthica* (Babczynska et al., 2006). So, if we consider the acute exposition of pesticides on two species of spiders named *P. lugubris* and *A. labyrinthica*, the survival fitness of *P. lugubris* would get privilege over *A. labyrinthica,* and its mode of diversification would not only ensure its survival in the stiff challenging environmental condition, it would most likely promote this species to get a long span of life-line in the course of evolution journey of life.

However, it is a growing concern among the researchers in the domain of molecular biology to realize that the presence of chemicals in the diverse grade of concentrations, its logistics through the ecosystems and its possession that is contained by the biological organism, and its cyclic rotation in the atmosphere and the biological organism would render the speciation practicable, even the presence of pollutants might be traced out at trace levels (Ure and Davidson, 2002). Deciphering the occurrence and elucidation of chemical speciation of any biological species is a cumbersome process, yet, the existence of a high level of chemical pollutants (either air pollutants like SO_2 and heavy metals like Pb) inside and in the ambiance of any organism would often alter the inherent molecular genetics and phylogenomics of the organism like spiders, to be evolved further in a diversified form, so undertaking of chemical speciation of spiders is the next level of challenges to the evolutionary biologists, that needs to be done in future (Hughes et al., 1980). It's quite interesting to witness that the changes in behavioral features of the same spider species vary from the changes in its habitat. To monitor the behavioral changes with the variations of adaptive changes and cognitive-behavioral changes among spiders, scientists have attempted to determine whether alteration of adaptive features would result in any new or altered characteristic features or traits of the concerned species in an irreversible way that might render the evolution of the new varieties or not. Arachnologists have coordinated such investigations on several jumping spiders, *Phidippus audax*, collected from wild habitats and many laboratory-bred jumping spiders (grown in small-sized cages with or without a painted dowel) of the same species. After a series of biobehavioral investigations, it has been noticed that the field-caught spiders are proactive than the laboratory-reared spiders to respond to the videotaped prey (Carducci and Jakob, 2000). Furthermore, field-caught specimens performed better in "detour tests" and activity in the open field than their counterparts reared in laboratory environments (Carducci and Jakob, 2000). So, these biobehavioral investigations rendered the researchers to make a precise observation that the alteration of adaptive conditions and the changing of living conditions of any species, which reared in different microclimatic conditions or different habitats, might alter the irreversible changes of biobehavioral traits of the same species, which might promote the evolution of ecotypes in future.

References

Anderson, J.F., Prestwich, K.N., 1982. Respiratory gas exchange in spiders. Physiol. Zool. 55, 72–90.

Andre, H., 1977. Introduction a l'etude ecologique des communautes de microarthropodes corticoles soumises a la pollution atmospherique. II. Recherche de bioindicateurs et d'indices biologique de pollution. Annls Soc. R. Zool. Belg. 106, 211–224.

Auld, J.R., Agrawal, A.A., Relyea, R.A., 2010. Re-evaluating the costs and limits of adaptive phenotypic plasticity. Proc. R. Soc. Ser. B 277, 503–511.

Babczyńska, A., Wilczek, G., Migula, P., 2006. Effects of dimethoate on spiders from metal pollution gradient. Sci. Total Environ. 370 (2–3), 352–359. https://doi.org/10.1016/j.scitotenv.2006.06.024.

Barnes, R.D., Barnes, B.M., 1955. The spider population of the abstract broomsedge community of the southeastern piedmont. Ecology 36, 658–666.

Bengtsson, G., Rundgren, S., 1984. Ground-living invertebrates in metal- polluted forest soils. Ambio 13, 29–33.
Borges, R.M., 2009. Phenotypic plasticity and longevity in plants and animals: cause and effect? J. Biosci. 34, 605–611.
Borgesen, L.W., Christensen, S.N., Clausen, I.H.S., Clausen, M.W., Hansen, O.C., Madsen, P.B., Rehder, S., 1984. En flora- og faunaundersogelse i Nordmarken, Laeso. A Report at the Institute of Population Biology. University of Copenhagen Denmark.
Botero, C., Weissing, F., Wright, J., Rubenstein, D., 2015. Evolutionary tipping points in the capacity to adapt to environmental change. Proc. Natl. Acad. Sci. U. S. A. 112, 184–189.
Bozinovic, F., Canals, M., 2008. Fisiología ecológica de mamíferos: compromisos y restricciones en el uso de la energía. In: Muñoz-Pedreros, A., Yañez, J.L. (Eds.), Mamíferos de Chile, second ed. Ediciones CEA, Valdivia, pp. 267–287.
Bozinovic, F., Contreras, L.C., Rosenmann, M., Torres-Mura, J.C., 1985. Bioenergética de Myotis chiloensis (Quiróptera: Vespertilionidae). Rev. Chil. Hist. Nat. 58, 39–45.
Canals, M., Rosenmann, M., Bozinovic, F., 1989. Energetics and geometry of huddling in small mammals. J. Theor. Biol. 141, 181–189.
Canals, M., Salazar, M.J., Durán, C., Figueroa, D., Veloso, C., 2008. Respiratory refinements in the mygalomorph spider *Grammostola rosea* Walckenaer 1837 (Araneae, Theraphosidae). J. Arachnol. 35, 481–486. https://doi.org/10.1636/JOASh06-47.1.
Canals, M., Figueroa, D., Alfaro, C., Kawamoto, T., Torres-Contreras, H., Sabat, P., Veloso, C., 2011. Effects of diet and water supply on energy intake and water loss in a mygalomorph spider in a fluctuating environment of the central Andes. J. Insect Physiol. 57, 1489–1494. https://doi.org/10.1016/j.jinsphys.2011.07.016.
Canals, M., Veloso, C., Solís, R., 2015. Adaptation of the spiders to the environment: the case of some Chilean species. Front. Physiol. 6, 220. https://doi.org/10.3389/fphys.2015.00220.
Carducci, J.P., Jakob, E.M., 2000. Rearing environment affects behaviour of jumping spiders. Anim. Behav. 59 (1), 39–46. https://doi.org/10.1006/anbe.1999.1282.
Clausen, I.H.S., 1984. Notes on impact of air pollution (SO_2 & Pb) on spider (Araneae) populations in North Zealand, Denmark. Ent. Meddr. 52, 33–39.
Clausen, I.H.S., 1986. The use of spiders (Araneae) as ecological indicators. Bull. Br. Arachnol. Soc. 7 (3), 83–86.
Cotto, O., Ronce, O., 2014. Maladaptation as a source of senescence in habitats variable in space and time. Evolution 68, 2481–2493.
Curtis, D.J., 1978. Community parameters of the ground layer araneid - opilionid taxocene of a Scottish island. Symp. Zool. Soc. Lond. 42, 149–159.
De Bont, R., Lachmund, J., 2017. Spatializing the History of Ecology: Sites, Journeys, Mappings. Routledge, Taylor & Francis Group, London and New York, p. 231.
De Vito, J., Formanowics, D.R.J.R., 2003. The effects of size, sex, and reproductive condition on thermal and desiccation stress in a riparian spider *Pirata sedentarius* (Araneae, Lycosidae). J. Arachnol. 31, 278–284. https://doi.org/10.1636/02-20.
De Vito, J., Meik, J., Geron, M., Formanowics, D.R.J.R., 2004. Physiological tolerances of three sympatric riparian wolf spiders (Araneae: Lycosidae) correspond with microhabitat distribution. Can. J. Zool. 82, 1119–1125. https://doi.org/10.1139/z04-090.
Demarino, N., 2017. The Importance of Spiders to an Ecosystem. https://animals.mom.me/importance-spiders-ecosystem-6242.html.
Dudley, R., 1998. Atmospheric oxygen, giant Paleozoic insects and the evolution of aerial locomotor performance. J. Exp. Biol. 201, 1043–1050.
Duffey, E., 1968. An ecological analysis of the spider fauna of sand dunes. J. Anim. Ecol. 37, 641–674.

Falkowski, P.G., Katz, M.E., Milligan, A.J., Fennel, K., Cramer, B.S., Aubry, M.P., Berner, R.A., Noacek, M.J., Zapol, W.A., 2005. The rise of oxygen over the past 205 million years and the evolution of large placental mammals. Science 309, 2202–2204. https://doi.org/10.1126/science.1116047.

Figueroa, D.P., Sabat, P., Torres-Contreras, H., Veloso, C., Canals, M., 2010. Participation of book lungs in evaporative water loss in *Paraphysa parvula*, a migalomorph spider from Chilean Andes. J. Insect Physiol. 56, 731–735. https://doi.org/10.1016/j.jinsphys.2010.01.001.

Foelix, R.F., 2011. Biology of Spiders, third ed. Oxford University Press, New York, p. 419.

Forsman, A., 2015. Rethinking phenotypic plasticity and its consequences for individuals, populations and species. Heredity 115, 276–284. https://doi.org/10.1038/hdy.2014.92.

Frick, H., Kropf, C., Nentwig, W., 2007. Laboratory temperature preferences of the wolf spider *Pardosa riparia* (Araneae, Lycosidae). Bull. Br. Arachnol. Soc. 14, 45–48. https://doi.org/10.13156/arac.2007.14.1.45.

Fusco, G., Minelli, A., 2010. Phenotypic plasticity in development and evolution: facts and concepts. Introduction. Phil. Trans. Roy. Soc. Lond. B Biol. Sci. 365 (1540), 547–556. https://doi.org/10.1098/rstb.2009.0267.

Gabriel, W., 2005. How stress selects for reversible phenotypic plasticity. J. Evol. Biol. 18 (2005), 873–883. https://doi.org/10.1111/j.1420-9101.2005.00959.x.

Gabritschevsky, E., 2005. Experiments on color changes and regeneration in the crab-spider, *Misumena vatia*. J. Exp. Zool. 47, 251–267. https://doi.org/10.1002/jez.1400470207.

Gilbert, O.L., 1971. Some indirect effects of air pollution on barkliving invertebrates. J. Appl. Ecol. 8, 77–84.

González-Lagos, C., Sol, D., Reader, S.M., 2010. Large-brained mammals live longer. J. Evol. Biol. 23, 1064–1074.

Greenstone, M.H., Bennett, A.F., 1980. Foraging strategy and metabolic rate in spiders. Ecology 61, 1255–1259. https://doi.org/10.2307/1936843.

Grossi, B., Canals, M., 2015. Energetics, scaling and sexual size dimorphism in spiders. Acta Biotheor. 63, 71–81. https://doi.org/10.1007/s10441-014-9237-5.

Hadley, N.F., Ahearn, G.A., Howarth, F.G., 1981. Water and metabolic relations of cave-adapted and epigean lycosid spiders in Hawaii. J. Arachnol. 9, 215–222.

Hanna, C.H.J., Cobb, V.A., 2007. Critical thermal maximum of the green lynx spider *Peucetia viridians* (Araneae, Oxyopidae). J. Arachnol. 35, 193–196. https://doi.org/10.1636/SH06-01.1.

Heublin, D., 1982. Untersuchungen zum EinfluB eines Waldrandes auf die epigaische Spinnenfauna eines angrenzenden Halbtrockenrasens. In: Hecken und Flurgeholze- Struktur, Funktion und Bewertung. Akademiefur Naturschutz und Landschaftspflege, Laufener Seminarbeitrage 5/82.

Hughes, M.K., Lepp, N.W., Phipps, D.A., 1980. Aerial heavy metal pollution and terrestrial ecosystems. Adv. Ecol. Res. 11, 217–372.

Ito, Y., 1964. Preliminary studies on the respiratory energy loss of a spider, Lycosa pseudoannulata. Res. Pop. Ecol. 6, 13–21. https://doi.org/10.1007/BF02524940.

Kokko, H., Sutherland, W.J., Johnstone, R.A., 2001. The logic of territory choice: implications for conservation and source-sink dynamics. Am. Nat. 157, 459–463.

Llandres, A.L., Figon, F., Christidès, J.P., Mandon, N., Casas, J., 2013. Environmental and hormonal factors controlling reversible colour change in crab spiders. J. Exp. Biol. 216 (Pt 20), 3886–3895. https://doi.org/10.1242/jeb.086470.

MacNaughton, R.B., Cole, J.M., Dalrymple, R.W., Braddy, S.J., Briggs, D.E.G., Lukie, T.D., 2002. First steps on land: arthropod trackways in Cambrian-Ordovician aeolian sandstone, southeastern Ontario, Canada. Geology 30, 391–394. https://doi.org/10.1130/0091-7613(2002)030<0391:FSOLAT>2.0.CO.

Marczyk, G., Migula, P., Trzcionka, E., 1993. Physiological responses of spiders to environmental pollution in the Silesian Region (Southern Poland). Sci. Total Environ. (2), 1315–1321. Elsevier Science Publishers B. V., Amsterdam.

Morse, D.H., 2007. Predator upon a Flower: Life History and Fitness in a Crab Spider. Harvard University Press, Cambridge, MA.

Moseholm, L., 1981. Biomonitering Afbly Omkring Accumulatorfabriken AIS LYAC, Iavundersogelse. Report from COWI consult, Teknikerbyen 45, DK-2830 Virum, Denmark.

Nentwig, W., 2013. Ecophysiology of Spiders. Springer-Verlag, Berlin, p. 529.

Panchanathan, K., Frankenhuis, W.E., 2016. The evolution of sensitive periods in a model of incremental development. Proc. R. Soc. B 283. https://doi.org/10.1098/rspb. Article 20152439.

Pigliucci, M., 2001. Phenotypic Plasticity: Beyond Nature and Nurture. Johns Hopkins University Press, Baltimore, MD, USA.

Pilegard, K., 1978. Airborne metals and SO_2 monitored by epiphytic lichens in an industrial area. Environ. Pollut. 17, 81–92.

Price, P.W., Rathcke, B.J., Gentry, D.A.G., 1974. Lead in terrestrial arthropods: evidence for biological concentration. Environ. Entomol. 3, 370–372.

Rabaud, E., 1919. Deuxieme note sur l'adaptation l'hochromie des Thomisides. Bull. Soc. Zool. France 53, 327–329.

Ratikainen, I.I., Kokko, H., 2019. The coevolution of lifespan and reversible plasticity. Nat. Commun. 10, 538. https://doi.org/10.1038/s41467- 019- 08502-9.

Robinson, J.V., 1981. The effect of architectural variation in habitat on spider community: an experimental field study. Ecology 62, 73–80.

Schaefer, M., 1978. Some experiments on the regulation of population density in the spider *Floronia bucculenta* (Araneae: Linyphiidae). Symp. Zool. Soc. Lond. 42, 203–210.

Schmidt-Nielsen, K., 1997. Animal Physiology. Adaptations and Environment, fifth ed. Cambridge University Press, New York, NY.

Shulz, S., Dunbar, R., 2010. Encephalization is not a universal macroevolutionary phenomenon in mammals but is associated with sociality. Proc. Natl. Acad. Sci. U. S. A. 107 (50), 21582–21586. https://doi.org/10.1073/pnas.1005246107.

Stevens, M., Merilaita, S., 2008. Animal camouflage: current issues and new perspectives. Phil. Trans. Roy. Soc. Lond. B Biol. Sci. 364, 423–427. https://doi.org/10.1098/rstb.2008.0217.

Tanaka, K., Ito, Y., 1982. Decrease in respiratory rate in a wolf spider, *Pardosa astrigera*, under starvation. Res. Popul. Ecol. (Kyoto) 24, 360–374. https://doi.org/10.1007/BF02515582.

Tufto, J., 2015. Genetic evolution, plasticity, and bet-hedging as adaptive responses to temporally autocorrelated fluctuating selection: a quantitative genetic model. Evol. Int. J. Org. Evol. 69, 2034–2049.

Uetz, G.W., 1989. The "ricochet effect" and prey capture in colonial spiders. Oecologia 81, 154–159. https://doi.org/10.1007/BF00379799.

Ure, A.M., Davidson, C.M. (Eds.), 2002. Introduction. In: Chemical Fractionation in the Environment. Blackwell Science, Oxford.

Varshney, C.K., Garg, J.K., Lauenroth, W.K., Heitschmidt, R.K., 1979. Plant responses to sulfur dioxide pollution. CRC Crit. Rev. Environ. Contr. 9 (1), 27–49. https://doi.org/10.1080/10643387909381667.

Veloso, C., 2003. Energética reproductiva de pequeños mamíferos. In: Bozinovic, F. (Ed.), Fisiología Ecológica y Evolutiva. Teoría y Casos de estudios en Animales. Ediciones Universidad Católica de Chile, Santiago, pp. 423–440.

Veloso, C., Bozinovic, F., 1993. Dietary and digestive constraints on basal energy metabolism in a small herbivorous rodent (*Octodon degus*). Ecology 74, 2003–2010. https://doi.org/10.2307/1940843.

Vlijm, L., Kessler, A., Richter, C.J.J., 1963. The life history of *Pardosa amentata* (Araneae, Lycosidae). Entomol Ber. Amsterdamm 3, 75–80.

Vollmer, A.T., Mac Mahon, J.A., 1974. Comparative water relations of five species of spiders from different habitats. Comp. Biochem. Physiol. 47, 753–765. https://doi.org/10.1016/0300-9629(74)90035-8.

Wade, K.J., Flanagan, J.T., Currie, A., Curtis, D.J., 1980. Roadside gradients of lead and zinc concentrations in surface-dwelling invertebrates. Environ. Pollut. 1, 87–93.

West-Eberhard, M.J., 1989. Phenotypic plasticity and the origins of diversity. Annu. Rev. Ecol. Systemat. 20 (1), 249–278.

Wikelski, M., Thom, C., 2000. Marine iguanas shrink to survive El Nino. Nature 403, 37–38.

Wilder, S.M., 2011. Spider nutrition: an integrative perspective. In: Simpson, S.J., Casas, J. (Eds.), Spider Physiology and Behavior: Physiology, Adv. Ins. Physiol., vol. 40, pp. 87–136. https://doi.org/10.1016/B978-0-12-387668-3.0000.

CHAPTER THIRTEEN

Biomonitoring of environmental pollution using spiderwebs

Abstract

Ecologists and environmental biologists have scientifically proven that spiders are one of the best bioindicators among arachnids. Scientists attribute the spider's long history of evolutionary success to spider silk, one of the best natural devices, and spiderwebs, considered "passive samplers, used mostly selective sorbents," which have helped environmental scientists analyze and retrieve microenvironmental changes wherever they occur. Spiders inhabit nearly every environment, and hence, diverse spiderwebs are easily accessible, costless samples that hardly change with the sunlight, temperature, rainfall, or wind. Furthermore, scientists have observed that spider silk can usually be sustained in polluted environments for long periods and absorbs toxins and dust particulates for the long term, which helps in biomonitoring the dynamics of the microenvironmental profile of any habitat.

Keywords: Arachnids; Biomonitoring; Environmental pollutants; Microevolution; Spiderwebs.

As these weather events [floods, fires, or hurricanes] can perturb natural communities in very dramatic ways, it will be important to understand which organisms are likely to persist or at least able to recolonize areas quickly. We noticed a large amount of dead vegetation at our site. Yet detritivores like millipedes and isopods that normally break down decomposing vegetation were largely gone … Instead, we recorded an increase in large-bodied groups like field crickets, spiders, and beetles. This change may have important consequences for the survival of amphibians, mammals, and reptiles, which opportunistically prey upon these larger invertebrate groups, or for birds as they fuel up for migration. Conversely, as there are now fewer invertebrates overall, groups that cannot reliably capture highly mobile invertebrates may be negatively impacted.

Karl Roeder

According to a health report on global population published by Institute for Health Metrics and Evaluation published in 2015, approximately four million people died worldwide from intermittent toxic exposure to environmental pollution, and the five key ailments were as follows (Health Effects Institute, 2017):

Spider Evolution
ISBN: 978-0-323-90041-6
https://doi.org/10.1016/B978-0-323-90041-6.00013-9

chronic obstructive pulmonary disease: 27.1%
ischemic heart disease: 17.1%
lower respiratory infections: 24.7%
lung cancer: 16.5%
stroke: 14.2%

There is no doubt that such potential health threats are primarily caused by aggressive industrial operations, followed by supplementary factors such as widespread urban sprawl, unsustainable agricultural practices and management of forest resources, drastic shrinkage of green cover, unplanned infrastructural development road transport systems, power generation, poor waste disposal management, etc. These factors have aggravated the surge in pollutants in the biosphere in the current regime of the Anthropocene. With issues such as the effect of pollutants, we are typically concerned about the human impact rather than acknowledging the base reality that such pollutants have an irreversible impact on our natural ecosystems, and the effect of pollutants on lower animals such as invertebrates (e.g., insects, spiders) has mostly remained an issue out of the scope of mainstream science. To understand the impact of pollutants in the environment, a unique mode of biomonitoring protocol was developed by environmental biologists at the beginning of the 1980s, when scientists postulated and refined the biological monitoring protocol with spiders (as a biological monitoring tool for appraising environmental pollution) for the first time (Rutkowski et al., 2018). The term biological monitoring was formally defined in a seminar held at Luxemburg organized by the National Institute for Occupational Safety and Health and the Occupational Safety and Health Organization that was cosponsored by European Economic Community. The main objective of this seminar was to determine the potential opportunities of biological monitoring as a multidisciplinary approach in protecting the health of workers continuously exposed to toxic pollutants as a part of their working environments. According to the earlier perception, biological monitoring was defined as the assessment or monitoring of "workplace agents (pollutants) or its derivatives, the metabolites examined in the tissues, excreta exhaled breathe or any combination of these that could help the medical and health professionals to assess the potential health risk of the workers, exposed to such 'workplace agents' or pollutants for long span" (World Health Organization, 1996), Biomonitoring is considered a complementary, alternative method to analyze and substantiate physical, chemical, and physiological data to track changes in the environment where large-scale, expensive, and extensive monitoring of environmental change is unlikely, and its impact on the biological organisms is difficult to measure. To achieve such an objective goal, a wide variety of biological species (e.g., lichens such as *Lobaria pulmonaria*, microorganisms such as *Geobacillus* sp., and a number of macro-organisms such as crayfish, toads, etc.) are used as bioindicators to monitor environmental undulations in concerned aquatic or terrestrial ecosystems (Lambert, 1997; Barbour et al., 1999; Protak Scientific, February 3, 2017). Ecologists and environmental biologists have noticed that spiders are one of the best bioindicators, belonging to arachnids, and their silk is one of the best natural devices for its evolutionary success, and they also considered "*passive samplers, used mostly selective sorbents*", as they are more potent than any other monitoring devices to analyze air quality using electronic, physicochemical devices competent for short-term monitoring of changing environment rather than long-term monitoring of air pollution. From an ethical perspective, for biomonitoring, the sample collection process is a noninvasive process where no living spiders are harmed or killed when collecting their webs (Hose et al., 2002; Xiao-Li et al., 2006; Rybak et al., 2012). As the spider inhabits almost every environment, spiderwebs are easily accessible and costless samples that do not change with changes in sunlight, temperature, rainfall, or wind, and scientists have observed that spider silk can be sustained in a polluted environments for a long time, and it absorbs toxins and dust particulates over the long term, which helps in monitoring the microenvironmental profile of any habitat over time (Rutkowski et al., 2018).

If we look at the functional aspects of spiders such as most common orb-weaving spiders, they are engaged in weaving spiderwebs using silk threads employing biophysical, biochemical, biobehavioral, and physiological maneuvering to backup foraging (i.e., trapping the flying insects and get nourishment from it) to ensure their survival. The greatest advantage of monitoring pollutants using spiderwebs is that these webs are electrostatically charged, allowing them to trap minute flying insects, pollens, spores, and other biological entities along with different organic and inorganic mineral particles (recognized as potential pollutants) of compatible electrostatically charged entities to spiderwebs and also provides an indicator of the concerned environment's health (Craig, 2019). Flying and air-blown objects, pick up the electrostatic charge in a way similar to that of the combs (made up of polyethylene compound) picked up the charge during moving and scrubbing through the hairs during a hair-dressing session or anyone shuffles across the carpets or rugs in her or his socks. The careful observation of scientists revealed that the silk-threaded spiderwebs stretched out and bends toward the charged flying insects or particles to trap it (Craig, 2019). Though the stretching potential of spiderwebs varies on the size of the object as well as the charge of it, it has a maximum range of 2 mm, and the maximum size of the object (or prey) ranges to 10 times larger in comparison with the thickness of the silk thread. Apart from the insects, the inorganic and organic, micro- and nanoparticles, trapped by the Orb-web spiders in their webs, range from toxic aerosols, fine-grained carbon deposits, heavy metals from the burning fuels, are harmful to the health of spiders themselves as they are not able to ingest the old webs or become sick after ingesting the old webs, containing the toxic air pollutants and that stalled spiders to make a new web (Craig, 2019). However, analyzing the chemicals, nonmetallic and metallic pollutants of spiderwebs, scientists predict the health of the concerned physical and biological environment. The comparative biochemical and toxicological account of the spider's web, collected from pollution-pronged and pollution-free regions would indicate the impact of specific pollutants on the health of the living organisms, specifically on human beings, living in the polluted environment that is the ultimate objective of biomonitoring.

It is required to mention here that many environmental factors impacted the web-spinning properties of the orb-web spiders and in turn, it renders the prey-catching potential of the concerned spiders. In a field-based ecological investigation, scientists observed that the Web-building spider, *Seothyra henscheli* (Eresidae), inhabiting the inhospitable dune ecosystems of the Namib desert has highly been influenced by environmental factors in terms of web design and foraging activities (Lubin and Henschel, 1996). The meticulous observation of scientists revealed that the wind-blown sands, which covered the webs to some extent, affect the capture of the number of prey and altered the extent of foraging of these psammophilous spiders, adapted to live in this dune ecosystem. The careful observation of Lubin and Henschel (1996) ascertained that the spider's habitat in the open and wind-exposed areas have a higher degree of sand-loading than the habitats in the sheltered area and the webs, and a sustained higher degree of wind current traps more small-sized prey than webs with low exposure to wind currents. Web-building and the maintenance of web construction under the high degree of challenges posed by undulating environmental conditions are found to be energetically expensive to *Seothyra henscheli* along with the potential threat of predation by potential predators of desert ecosystems, exposition of extreme heat and water-loss due to desiccation and the changing environmental factors, changing of seasonal cycles, etc.

in the ambiance of this psammophilous spider, rendered its web-spinning potential and the entrapping potential of insects, particulate substances, pollutants in such webs. So the short-term biomonitoring of toxins, pollutants from examining a spider's web might reveal an inconsistent result, whereas a persistent result of environmental health would supposedly give the consistent result of the microenvironment of any location.

For a long time it has been observed that in and around the urban and industrial human settlements, the heavy metals and nonbiodegradable compounds of carcinogenic potentials, the toxic pollutants, the emissions of anthropogenic activities (due to exponential rate of industrial operations, a surge of vehicle operations, etc.), environmental pollutants such as As (Arsenic), Ag (Silver), Pb (lead), Cd (cadmium), Cr (Chromium), Cu (copper), Mn (manganese), Ni (Nickel) was integrated to the trophic chains of the concerned ecosystems (Rutkowski et al., 2018). On the basis of qualitative experimental investigations, environmental scientists, toxicologists, and arachnologists have noticed that spiderwebs weaved by the species belonging to the families of Agelenidae, Pholcidae, and Linyphiidae are excellent bioindicators of environmental pollutants (Rybak et al., 2012; Rybak and Olejniczak, 2014; Rybak, 2015).

Environmental scientists observed that the by-products of industrial processes and incomplete incineration of the municipal wastes, trashes and the burning of medical wastes yield Dioxins, environmental pollutants get integrated with the fatty tissues of animals and it has immense mutagenic and carcinogenic potential (Rutkowski et al., 2018). Scientists observed that environments exposed to such toxic compounds as polychlorinated dibenzo-para-dioxins (PCDDs) absorbed by spiderwebs inhabited by spiders in such a polluted environment. The biochemical monitoring of Rybak and Rutkowski (2018) revealed that the spiderwebs in the polluted environment of Lower Silesia, Poland, has absorbed more PCDDs closed by the medical waste incinerator and the nearby location of the municipal waste disposal/burning location. Such observation indicated further that the high index of PCDDs in a certain location, beyond the threshold limit, might aggravate the mutagenesis or carcinogenesis among a number of animal groups and it might render human-induced microevolutionary changes or anthropogenic extinction of life in a certain direction in the polluted environment.

Very often scientists affirmed that chemical analysis of air samples is not enough to understand its cholinergic interactions among interactive individual compounds, due to excessive pharmacodynamics response and possible toxic response on biological organisms (Claxton et al., 2004). To understand the biological impact of chemical pollutants, bacterial test for mutagenicity, a standard analyzing protocol, which has been proposed by Ames et al. (1973) and popularly recognized as "Ames Mutagenicity Assessment," randomly used nowadays. As the arachnologists and environmental biologists have noticed that spiders belong to Pholcidae and Agelenidae usually grow in human settlements and the outdoor environments such as road tunnels, bridges, etc., where spiders inhabited for a whole year and weave the spiderwebs could be a good source of a long-term monitoring tool for assessment of air pollution in a location. Rybak et al. (2017) and Rutkowski et al. (2018) have successfully used the spider's web to coordinate "Ames Mutagenicity Assessment" to analyze interior and exterior air pollution. Rybak et al. (2017) have noticed further that the samples of spider's web collected from the interior environment, exposed to petrogenic and pyrogenic emissions (such as garage room, a room adjacent to high traffic locations and the boiler and furnace room, etc.), analyzed to be respondent in the mutagenic activity.

The contemporary investigations of environmental biologists, molecular biologists, and toxicologists revealed polycyclic aromatic hydrocarbons (PAHs) and their derivatives found in coal and tar deposits and formed of incomplete combustion or pyrolysis of organic material (from engines, incinerators, and biomass burning of forest fires, etc.). These toxic pollutants appeared as potential mutagenic and carcinogenic threats to human health and pose serious health issues for other mammals, birds, etc. (Widziewicz et al., 2017). Rybak and Olejniczak (2014) have used a spider's web (of Agelenidae) sample to analyze the excessively high PAH level in the polluted ambiance of Wroclaw, Poland, the high traffic urban area having a great number of vehicles with diesel engines. Scientists noticed further that spiderwebs in such location where air pollutants like PAH act as the pollution indicator for the long term that might alter and enhance the rate of irreversible mutagenesis among the living organism in its vicinity and it might alter the process of evolution or extinction of the concerned organism in a long run.

To identify the various inorganic particles as well as pollutants, either deposited or absorbed by the spiderwebs have scientifically analyzed and examined by adopting a unique bioanalytical technique, knowing as "Scanning Electron Microscope with Energy Dispersive X-ray Analyser" (Gorka et al., 2018). Scientists observed that this unique technique has been found effective in determining the morphological, chemical, and geological identity of the trace elements, found on the web surface. It has further been observed by environmental biologists that the identity of the trace elements whose particles are absorbed by the spiderwebs varies, based on the nature of the webs, weaved by the specific species of the spider (Such as the webs spun by species belonging to Pholcidae are found to be more competent to precisely trap the particulate air pollutants and minerals of any specific environment better than the members of Agelenidae) the time of exposition and the nature of toxicity of the pollutants present in the air (Rutkowski et al., 2018). Scientists also observed that factors like "mineralogical characteristics of atmospheric dust," and their "chemical and isotopic signature," etc. also played a key role in the virtual representation of the pollutants in the sample of spider's web, examined in the laboratory condition.

The biochemical analysis of the toxic and heavy metals found in spiderwebs, collected from three caves (Jenolan, Abercrombie, and Wombeyan caves) of the South Wales region of Australia, shows the excessive level of Pb and Zn in the microenvironment of Jenolan cave and environmental scientists recognized that "vehicular traffic emission" went through the arch of it, accumulated by the local spiderwebs for time (Hose et al., 2002). Xiao-li et al. (2006) have conducted the level of environmental pollutants such as Pb, Zn, Cd, and Cu in the air polluted with excessive level of vehicular traffic emission sites in China using the web sample analysis of two common spiders—*Araneus ventricosus* and *Achaearanea tepidariorum*. Rybak et al. (2012) have investigated emissions of pollutants released from vehicles in Wroclaw, Lower Silesia, Poland, by analyzing spiderwebs of the two species of Agelenidae (*Malthonica silvestris* and *M. ferruginea*), which inhabit road tunnels, bridges, and parking lots of urban ecosystems of Poland. According to principal components analysis (PCA), on *M. silvestris* webs, the top 10 metallic pollutants in the descending order are as follows (Rybak, 2015): Fe (iron)> K (potassium)> Al (aluminum)> Zn (zinc)> Mg (magnesium)> Ti (titanium)> Mn (manganese)> Cu (copper)> Pb (lead)> Ni (nickel).

At the next level, scientists have tried to carry out a comparative evaluation of the presence of Pb and Zn in the industrial and urban environment of Wroclaw, Poland, as per PCA analysis on the web sample of *M. silvestris* and the particulate matter index (PMI) of bioaerosol of this location scientists observed that PCA index of the heavy metal content of any location, did not necessarily correspond to the content of submicron particles or PMI index as scientists have contemplated that the PMI index is of instantaneous reading of the pollutants of any microenvironment, whereas the PCA index is the cumulative absorption of the pollutants by the spider index for a long time (Rutkowski et al., 2018). Whatever the result of discrepancy persisted between PCA or PMI results, the PCA index has been considered a qualitative analytical index of the heavy metal pollution in an extremely polluted environment.

Environmental biologists have observed that the exposure of pollutants, specifically the metallic pollutants such as ferromagnetic particulate matter (PM) enhanced the magnetic susceptibility of the spider silk and the examination of the magnetic susceptibility of ferromagnetic PM in such samples of spiderwebs helped in understanding the pollutants under the conditions of any polluted environment. The application of such investigating protocol has been adopted by Rachwal et al. (2018) when he collected the samples of spider's web from the rural and urban areas in Poland, and the spider's web, collected and reared from the pollution-free lab, isolated the ferromagnetic PM and analyze its magnetic susceptibility and noticed that the presence of ferromagnetic pollutants in the polluted environment is substantially high, which is presumably generated by anthropogenic actions and the long-term exposition of metallic particulate matters, specifically the intermittent exposition of polluted air with ferromagnetic particle matters have enormous health issues (such as enhancing the level of neurotoxicity to create brain disorder order or cardiac failure in mammals) on the organisms inhabiting in such environments (Hounslow and Maher, 1996; Wright and Ding, 2016).

In this current regime of Anthropocene, we are living in the undulating environment of a biosphere, where climate change not only impacted the equilibrium between living and nonliving entities but also appeared as the key factor defining and redefining the sustainability of the living organisms by tilting the equilibrium between its mode of extinction or mode of evolutionary diversification. In the contemporary climate change regime of Anthropocene, we are experiencing frequent occurrence of natural disasters such as cascade actions of sweltering heatwaves, torrential downpours, smoky skies, toxic waters, etc. in different parts of our biosphere and all of these natural disasters, solely or collectively jeopardize the existence of living entities of our biosphere, partly due to climate change (Heyck-Willams, 2019). According to the recent scientific communications (Cook et al., 2013) and report of the Intergovernmental Science-Policy Platform on Biodiversity and Ecosystem Services (IPBES, 2019), the universal consensus among environmental scientists to acknowledge that the contemporary anthropogenic actions are responsible for radical climate change, furthermore, scientists appraised that climate change with other factors rendered around one million plants and animals in extinction throughout the world.

According to the conventional definition, a natural disaster is a key adverse event, resulting from natural processes of Earth and it would appear to be in any form ranges from floods, earthquakes, tsunamis, volcanic eruptions, tornadoes, hurricanes, etc. Scientists considered that an adverse event would not rise to that level where we could recognize it as a natural disaster if it occurs in such an area, which has not been inhabited by the vulnerable populations (Alexander et al., 2002; Wisner et al., 2003). Every natural disaster has a huge impact considering the loss of biodiversity in the land around the vicinity of the epicenter of such catastrophic environmental matrix and it is difficult to evaluate as well as substantiate the loss of biodiversity in terms of dollar value due to the occurrence of such natural disaster, driven by climatic change. In 2019, NOAA National Centers for Environmental Information published a report titled "U.S. Billion-Dollar Weather and Climate Disasters," in which it mentioned, "By October, there were 10 weather and climate disaster events in 2019 that each exceeded $1 billion in losses" (NOAA, 2019). According to the "Toba catastrophe hypothesis," geologists, paleobiologists, and evolutionary biologists estimate that around 75,000−80,000 years ago, a supervolcanic eruption (the deadliest volcanic eruption, grade 8) occurred at Lake Toba in Sumatra (Indonesia) and left only 10,000 human occupants, and many scientists have argued for a different figure of 1000 pairs of human breeding pairs, as the severity of this catastrophe put the entire human population of this location on the brink of extinction and created a bottleneck in human evolution (Gibbons, 2010). According to recent observations by the Intergovernmental Panel on Climate Changes, the current emission level of global greenhouse gas must be cut in half by 2030 and reach zero levels by next century to avoid the catastrophic impact of climate change; otherwise, 74% of the global human population would be exposed to a lethal heatwave that would wipe out much of that population (Lavelle, 2017; IPCC, 2018). In fact, natural disasters such as forest fires, floods, and hurricanes have an enormous and visible impact on human communities and jeopardize natural ecosystems, too. When the equilibrium state of natural communities has been perturbed by the occurrence of such natural disasters, it is essential to consider which organisms are integral to such ecosystems and would persist when facing such catastrophes and are able to recolonize the ecosystem to restore it. Scientists are convinced that if there are not enough such capable species in such fragile ecosystems to perform basic ecorestoration tasks like accelerated decomposition, streamline soil recycling, expedited pollination, and fertilization processes, reinstating such ecosystems to their earlier states would be unlikely (Floodlist News, 2018).

In 2007, marine biologists have investigated the cause of death of dozens of sea otters in California and they have reported that the possible cause of the death of the sea otters was the toxins, produced from an algal bloom in the fresh-water bodies, spilled into the oceans and entered the bodies of shellfish, which further was consumed by sea otters as a primary food source and were killed (CDC, 2018). In 2016, scientists reported that the algal bloom, known as "Red Tide," caused an enormous economic loss in the shellfish industry in the state of Massachusetts around $16 million loss (Anderson et al., 2016). In 2018, a havoc loss of marine biodiversity has been reported by marine biologists due to the worst algal bloom in Florida, the neurotoxins, spilled by the marine algae, found to kill precious marine species such as Kemp's Ridley turtles, whale sharks, dolphins, goliath grouper fish, etc. (Furby, 2018). Preliminary survey documentation of US Fish and Wildlife Services indicated around 20 narrowly distributed endemic species, having impressive population sizes, found in California, USA, before the occurrence of Hurricane "Irma" in 2017, and according to their post-Irma survey, around 14%–22% (around 1000 number of individuals) of the Key deer (*Odocoileus virginianus clavium*) population size, the smallest, locally endemic deer population in North America, has drastically shrunk, which might have created a bottleneck effect in the future (Nobel, 2017). The NOAA's *National Climate Report* mentioned that the sweltering heat waves in Alaska, which was around 7.9°C in 2019, rendered the extreme saturation of oxygen in the major water bodies in Alaska, causing acute suffocation, leading to the die-off of a substantial volume of Alaskan salmons (sockeye, pink, and chum varieties) (Berwyn, 11th July 2019; Montgomery, 2019; NOAA, 2019). Environmental biologists observed that frequent occurrences of wildfires and drastic human-induced climate change not only destroy wild forests but also lead to the wiping out of the pristine wildlife resources of those forest habitats. In 2016, the Soberanes Fire burnt down 130,000 acres of pristine green bush, and the economic loss of forest products in California, USA, from that fire was around $260 million, while the lost biodiversity cost the extinction of amphibians such as red-legged frogs and a substantial loss of trout populations because of the choking of water bodies with accumulation of burnt ash, carbon particles in water bodies, excessive sediment levels, etc. (Elliott, 2017). According to the impact assessment report of Hurricane Andrew, in 1992, in South Louisiana's Atchafalaya

Basin, USA, scientists have specifically noted the death of an estimated 184 million fish (DiLonardo, 2018). Based on their observation and analysis of the water sample of Atchafalaya Basin, Louisiana, scientists concluded that the ripped off tree branches, shrubs, tree trunks, the pile of storm-damaged foliage tossed up in the adjacent wetlands and carried away by flooded water and chocked the waterways of the shallow basin. The piled-up organic debris, heaped up at the bottom layer of the basin, decayed after receding the water level and putrefied organic waste in the basin rendered the level of dissolved oxygen in the water bodies very low and suffocated the school of fish in the Atchafalaya basin in South Louisiana, USA (DiLonardo, 2018). In 2015, a collaborative investigation by scientists from the University of Oklahoma and Cameron University, USA, the researchers under the leadership of Karl A. Roeder has originally planned to coordinate with the ecological effect of the controlled blaze of a piece of a wild habitat adjacent to the University of Oklahoma's biological station at Lake Texoma and a preburn sample survey of insects and other invertebrates such as spiders, millipedes, snails, etc. were made in March 2015 to compare the data with the postfire session. But a torrential rain in June 2015 led to a torrential flood in the lake and adjacent regions when the flood level rose to the water level of to 30 feet high and was considered an unprecedented flood disaster in the last 100 years history in South-Central Oklahoma, the USA in 2015. It paved a new opportunity for scientists to study the impact of the flood on the population dynamics especially the lower group of animals, insects, and arthropods in that area. The preliminary investigative comparisons in pre- and postflood seasons, yield a unique result as it shows there was a 93% decrease in the population of the lower group of invertebrates in that area, with a rate of 60% in decreasing number of species, whereas the decreasing rate of biomass found to be around 64% and the expression of Karl Roeder was notable in this regard, who said, *"The area had transitioned from a diverse insect community to a large expanse of crickets and spiders. We were pretty surprised at how similar everything was"* (Entomological Society of America, 2018). So it has been noticed that natural calamities like floods have ripple effect that not only render changes in the composition and functionality of any ecosystem but also expedite serial succession of its animals, and it has also been noticed that when the displaced species failed to recolonize or reassemble, the unique pace of serial succession sets into widen-up the scope of occupying the unoccupied ecological niches by new groups of living organisms that might either winding up the extinction of any extinction process or beginning of an evolutionary journey of another living being. But scientists did not have much strong evidence that would explain further about the animal that has the potential to turn around the aftermath of the natural catastrophes and competent enough to recolonize in the disturbed habitat and which one failed to do so. To elucidate the post-natural catastrophic effect as well as to draw a comparative account of recolonizing potential of some species in comparison with the challenge of poor sustainability issues, which has been witnessed among the extinction pronged species, a brief narration on the occurrence of disastrous Hurricane Florence in North Carolina, USA, in 2018 must be done here. According to the official record, around 48 people, across the three states around North Carolina, were killed along with a huge loss of livestock, such as the 3.4 million domestic fowl and 5500 captive hogs killed by the disastrous Hurricane Florence (Mancini, 2018). But environmental biologists have noticed that the wild herds of feral horses, which lived on the Tarheel State Barrier Islands in North Carolina, the USA, supposedly descended from mustangs and reportedly brought in by Spanish explorers in the USA. The brief survey aftermath of Hurricane Florence in North Carolina revealed that a large number of these beach—loving herds of feral horses were storm-savvy, and they sustained the catastrophic storm of Hurricane Florence whereas several other animals in the same wild habitat were found to have perished. The evolutionary biologists observed that these wild horses had sustained such vortex of severe storms for hundreds of years and they were accustomed to changes in the weather in advance so that they could move to the higher elevation and used to take refuge in the dense vegetated locations onset of hurricane, and this strategy has rendered the wild herd of the horses to sustain natural calamities (Mancini, 2018). At the same time, scientists have noticed that employing the temporary measure of moving out from its typical habitat to shift to its temporary refuge is most unlikely for the creatures such as crabs, oysters, shrimps, etc., that inhabit those coastal areas. The critical observations of marine biologists and environmental biologists revealed that the flood water, generated by hurricanes and tropical storms gushed in the estuaries and bays, brought down the salinity level, made the marine habitat unsuitable for the oysters and shrimps and many marine organisms (Mancini, 2018). Marine biologists have observed that the sharks, specifically blacktip sharks that could detect waterborne vibrations, caused by hurricane and sense the drop of barometric pressure in the water column left the shallow bays to move into the deep offshore water zone for securing its safety from the catastrophic vortex of such tropical storms in the surface region of shallow-water bodies, and when the storm ends they come back to the shallow bays. This going deep technique has successfully been adopted by the sharks but not been employed by the surface-breathing marine mammals such as dolphins or surface-breathing reptiles such as alligators, and it jeopardized the sustainability of these surface-breathing marine animals to survive in the storm-induced damages, at the shallow surface of the water bodies in the bay water areas (Mancini, 2018).

Environmental biologists noticed that the catastrophic effect of tropical cyclones in the United States and the Caribbean directly wrecked-up the forest resources and indirectly perished the wildlife resources, even expedite the extinction of many endangered or endemic species of these locations, the living of which is integrated to the existence of these forest ecosystems. According to a disaster evaluation report, prepared on the aftermath of Hurricane Hugo in South Carolina, USA, the surveyors reported massive damage to the forest ecosystems, vegetated by pine forests and the massive damage of the pine forests aggravated the loss of the largest population of red-cockaded woodpeckers, of the endangered avian species, dependent upon the existence of these pine forest ecosystems (Mancini, 2018). According to the "Tropical Cyclone Report—Hurricane Irene, Aug. 21–28, 2011," prepared by Avila and Cangialosi (2011), was one of the major tropical cyclones that wrecked the far and wide of the Caribbean and Eastern coast of the United States, which cost around $14.2 billion loss of United States, Caribbean, and Canada in terms of economic loss, which was the costliest natural disaster diminished a vast area of forest cover by uprooting, felling, and damaging centuries-old forests. The wildlife rehabilitation workers observed that as a part of collateral damage, wildlife lost its sustainability, such as smaller mammals like squirrels—the gusty wind strip off the fruits and nuts of the trees that caused the squirrels to struggle and suffer without food. It blew away a great number of baby squirrels or newborns of their arboreal nests that would ponder them to fend themselves in the chaotic aftermath of Irene if left unattended (Mancini, 2018). But it is not necessarily the general rule of thumb that all wildlife species respond to tropical storms and hurricanes in the same way as the wildlife biologists and field ecologists have observed that white throat sparrows have been evolved with sensing potential to adjudge dropping of barometric pressure, which stop them to postpone their plan of the long flight of migration during fall or in spring to avert the catastrophe of tropical storms aftermath (Mancini, 2018). According to another observation of the animal biologists and environmental biologists, made after the occurrence of Hurricane

Florence during the end of August-middle of September 2018 in North Carolina and Virginia, United States, it has been revealed that there was a surge in the population of spadefoot toad, a diminishing amphibian species in North Carolina and that happened immediately after the Hurricane Florence (Mancini, 2018). Scientists have further defined that the spadefoot toads preferred to breed and pass the juvenile stage in temporary waterbodies to avoid the predatory pressure of fish and other aquatic insects, reptiles, living in permanent water bodies such as lakes, streams, or ponds. As the stagnant flood water had rendered the low-level depressions like makeshift waterbodies but lack of predators in it, it became the ideal breeding ground for this spadefoot toads and that led to the population surge of this amphibian species in the backdrop of the intermittent threat of shrinkage of amphibian population, worldwide.

To move further, the evolutionary biologists contemplated that the tropical storms and hurricanes pushed the reptiles forward to get evolved with a strong grip and they have referred to the Anoles, a group of New World lizards who climb the walls "with adhesive toe pads" (Mancini, 2018). A meticulous evolutionary study on the evolution of "adhesive toe pads" in reptiles in 2018, has made the evolutionary biologists contemplate that in course of evolution maneuvered by the intermittent exposure of severe tropical storms and hurricanes, a reptilian species supposedly evolve with shorter hind legs and longer toe pads that made them fit to survive in clinging tree limbs to stick around their habitat to counteract the blowing vortex of strong wind current, generated onset of the hurricane and strong tropical storms (Mancini, 2018).

Amy Savage, one of the leading faculty of the Biology department in Rutgers University, USA had conducted a unique study along with her five coresearchers since 2012 and published her findings, titled "Homogenizing an Urban habitat mosaic: Arthropod diversity declines in New York City parks after Super Storm Sandy" in the journal of "Ecological Applications" in December 27, 2017. In this simple but unique investigation of Savage and her colleagues, it has been observed that the diversity of arthropods (comprising ants, bees, beetles, spiders, wasps, etc.) was considerably higher in the parks than the adjacent street medians, which went down after storm were almost indistinguishable from the street medians in the poststorm survey (Rutgers University, 2018). With the help of this unique investigation, scientists hypothesized that the organisms inhabiting the high-stress "urban street medians" are well adapted to such catastrophic disturbance and acquired behavioral adaptations rendering them more resilient to sustain natural disasters as well as catastrophic consequences than their counterparts living in habitats in lesser-stress environments like "city parks." Based on such findings, Savage said *"These data suggest that one possible result of the increasing frequency and intensity of extreme weather events will be homogenization of diversity in cities and that the direction of this simplification of urban communities may be quite predictable"* and she further commented *"Testing these contrasting hypotheses was an opportunity to not only help people understand and plan for diversity changes after extreme weather events, but also to provide important data that would move the field of ecology forward"* (Rutgers University, 2018).

References

Alexander, G.R., Kogan, M.D., Nabukera, S., 2002. Racial differences in prenatal care use in the United States: are disparities decreasing? Am. J. Pub. Health 92 (12), 1970–1975.

Ames, B.N., 1973. Carcinogens are mutagens: their detection and classification. Environ. Health Perspect. 6, 115–118.

Anderson, J.L., Anderson, C.M., Chu, J., Meredith, J., 2016. Fishery Performance Indicators Manual (Version 1.3). http://isfs.institute.ifas.ufl.edu/projects/new-metrics/fpi-manual.

Avila, L.A., Cangialosi, J., December 14, 2011. Tropical Cyclone Report Hurricane Irene (AL092011) 21–28 August 2011. www.nhc.noaa.gov/data/tcr/AL092011_Irene.pdf.

Barbour, M.T., Gerritsen, J., Stribling, J.B., 1999. Rapid Bioassessment Protocols for Use in Streams and Wadeable Rivers: Periphyton, Benthic Macroinvertebrates and Fish, second ed. U.S. Environmental Protection Agency (EPA). EPA 841-B-99-002, Washington, D.C. (Report).

Berwyn, B., July 11, 2019. Alaska Chokes on Wildfires as Heat Waves Dry Out the Arctic. https://insideclimatenews.org/news/11072019/arctic-wildfires-alaska-climate-change-heat-wave-2019-university-funding/.

CDC[Centers for Disease Control and Prevention], February 6, 2018. What Has Toxoplasma in Sea Otters Taught Us about the Risks and the One Health Approach to Global Public Health. https://emergency.cdc.gov/coca/calls/2018/callinfo_020618.asp.

Claxton, L.D., Matthews, P.P., Warren, S.H., 2004. The genotoxicity of ambient outdoor air, a review: Salmonella mutagenicity. Mutat. Res. 567 (2- 3), 347–399. https://doi.org/10.1016/j.mrrev.2004.08.002.

Cook, J., Nuccitelli, D., Green, S.A., Richardson, M., Winkler, B., Painting, R., Way, R., Peter Jacobs, P., Skuce, A., 2013. Quantifying the consensus on anthropogenic global warming in the scientific literature. Environ. Res. Lett. 8 (2). http://iopscience.iop.org/article/10. 1088/1748-9326/8/2/024024/meta.

Craig, T., June 2, 2019. How Spider Webs Can Detect Air Pollution? Broadcast 6.2 & 6.7.2019. https://www.montananaturalist.org/blog-post/how-spider-webs-can-detect-air-pollution/.

DiLonardo, M.J., September 12, 2018. What Happens to Animals during a Hurricane? https://www.treehugger.com/what-happens-wildlife-during- hurricane-4863881.

Elliott, S., February 21, 2017. Woodland Buffers Protect Amphibians, Ecosystems. https://www.usda.gov/media/blog/2016/02/24/woodland- buffers-protect-amphibians-ecosystems.

Entomological Society of America., March 15 , 2018. When Natural Disaster Strikes, Can Insects and Other Invertebrates Recover? 93 Percent Decline in Local Invertebrate Abundance after 2015 Flood. ScienceDaily. http://www.sciencedaily.com/releases/2018/03/180315140702.htm.

Floodlist News, March 16, 2018. Study Find Massive Decline in Insect Population after Floods. http://floodlist.com/dealing-with-floods/study- massive-decline-insects-after-floods.

Furby, K., August 17, 2018. Hundreds of Animals Are Dying in Red Tide. These People Are Trying to Save Them. www.washingtonpost.com/news/animalia/wp/2018/08/17/hundreds-of-animals-are-dying-in-red-tide-these- people-are-trying-to-save-them/.
Gibbons, A., January 19, 2010. Human Ancestors Were an Endangered Species. https://www.sciencemag.org/news/2010/01/human-ancestors- were-endangered-species#.
Gorka, M., Bartz, W., Rybak, J., 2018. The mineralogical interpretation of particulate matter deposited on Agelenidae and Pholcidae spider webs in the city of Wrocław (SWPoland): a preliminary case study. J. Aerosol Sci. 123, 63–75. https://doi.org/10.1016/j.jaerosci.2018.06.008.
Health Effects Institute, 2017. State of Global Air/2017: A Special Report on Global Exposure to Air Pollution and its Disease Burden. https://www.stateofglobalair.org/sites/default/files/soga_2017_report.pdf.
Heyck-Willams, S., 2019. Climate Change, Natural Disasters, and Wildlife. https://www.nwf.org/-/media/Documents/PDFs/Environmental- Threats/Climate-Change-Natural-Disasters-fact-sheet.ashx.
Hose, G.C., James, J.M., Gray, M.R., 2002. Spider webs as environmental indicators. Environ. Pollut. 120, 725–733.
Hounslow, M.W., Maher, B.A., 1996. Quantitative extraction and analysis of carriers of magnetization in sediments. Geophys. J. Int. 124 (1), 57–74. https://doi.org/10.1111/j.1365-246X.1996.tb06352.x.
IPBES, 2019. Summary for Policymakers of the Global Assessment Report on Biodiversity and Ecosystem Services of the Intergovernmental Science- Policy Platform on Biodiversity and Ecosystem Services. https://www.ipbes.net/global-assessment-report-biodiversity-ecosystem- services.
IPCC, 2018. Global Warming of 1.5°C. https://www.ipcc.ch/sr15/.
Lambert, M.R.K., 1997. Environmental effects of heavy spillage from a destroyed pesticide store near Hargeisa (Somaliland) assessed during the dry season, using reptiles and Amphibians as bioindicators. Arch. Environ. Contam. Toxicol. 32 (1), 80–93. https://doi.org/10.1007/s002449900158.
Lavelle, M., 2017. Deadly Heat Waves Could Endanger 74% of Mankind by 2100, Study Says. Inside Climate News. https://insideclimatenews.Org/news/19062017/heat-waves-world-populationrisk-endangered-climate-change-study.
Lubin, Y., Henschel, J., 1996. The influence of food supply on foraging behaviour in a desert spider. Oecologia 105, 64–73. https://doi.org/10.1007/BF00328792.
Mancini, M., October 1, 2018. How Do Animals Survive Major Storms? https://animals.howstuffworks.com/animal-facts/do-animals-survive- major-storms.htm.
Montgomery, M., October 10, 2019. Fish Farming Controversy over Massive Salmon Die-Off. https://www.rcinet.ca/en/2019/10/10/fish-farming- controversy-over-massive-salmon-die-off/.
NOAA National Centers for Environmental Information (NCEI), 2019. U.S. Billion Dollar Weather and Climate Disasters. https://www.ncdc.noaa.gov/billions/.
Nobel, J., October 30, 2017. Recent Hurricanes Pushed Rare Island Species Closer to the Brink. www.nationalgeographic.com/news/2017/10/hurricane-irma-hurts-florida-keys-wildlife.
Protak Scientific, February 3, 2017. What Are Biological Indicators? https://www.protakscientific.com/what-are-biological-indicators/.
Rachwał, M., Rybak, J., Rogula-Kozłowska, W., 2018. Magnetic susceptibility of spider webs as a proxy of airborne metal pollution. Environ. Pollut. 234, 543–551. https://doi.org/10.1016/j.envpol.2017.11.088.
Rutgers University, January 11, 2018. Urban Insects Are More Resilient in Extreme Weather: Rutgers University-Camden Rersearch Examins How Arthropods Survive Hurricanes in Urban Centers. www.sciencedaily.com/releases/2018/01/180111115332.htm.

Rutkowski, R., Rybak, J., Mach, T., Rogula–Kozłowska, W., 2018. Spider webs in monitoring of air pollution. SHS Web Conf. 57, 02011. https://doi.org/10.1051/shsconf/20185702011 (2018).

Rybak, J., 2015. Accumulation of major and trace elements in spider webs. Water Air Soil Pollut. 226, 105. https://doi.org/10.1007/s11270-015-2369-7.

Rybak, J., Olejniczak, T., 2014. Accumulation of polycyclicaromatic hydrocarbons (PAHs) on the spider webs in thevicinity of road traffic emissions. Environ. Sci. Pollut. Control Ser. https://doi.org/10.1007/s11356-013-2092-0.

Rybak, J., Rutkowski, R., 2018. Determination of PCDDs in spider webs: preliminary studies. E3S Web Conf. 28, 01033. https://doi.org/10.1051/e3sconf/20182801033 (2018).

Rybak, J., Sowka, I.M., Anna, Z., 2012. Preliminary assessment of use of Spider webs for the indication of air contaminants. Environ. Protect. Eng. 38 (3), 175–181.

Rybak, J., Hanus-Lorenz, B.E., Rogula-Kozłowska, W., Loska, K., Widziewicz, K., Rutkowski, R., 2017. Clean energy clean water and clean air. In: Kaźmierczak, B. (Ed.), International Conference on Advances in Energy Systems and Environmental Engineering (ASEE17), Book of Abstracts/, Wrocław: Oficyna Wydawnicza Politechniki Wrocławskiej, pp. 121–122.

Savage, A.M., Youngsteadt, E., Ernst, A.F., Powers, S.A., Dunn, R.R., Frank, S.D., 2017. Homogenizing an urban habitat mosaic: arthropod diversity declines in New York City parks after Super Storm Sandy. Ecol. Appl. 28 (1), 225–236. https://doi.org/10.1002/eap.1643.

WHO, 1996. Biological Monitoring of Chemical Exposure in the Work Place: Guidelines, vol. 1. Geneva. https://apps.who.int/iris/bitstream/handle/41856/WHO_HPR_OCH_96.1.pdf;jsessionid=549A78735B7B7498D8226AA1BA 3DB63F?sequence=1.

Widziewicz, K., Rogula-Kozłowska, W., Majewski, G., 2017. Lung cancer risk associated with exposure to benzo(A)pyrene in polish agglomerations, cities, and other areas. Int. J. Environ. Res. 11, 685–693. https://doi.org/10.1007/s41742-017-0061-z.

Wisner, B., Blaikie, P., Cannon, T., Davis, I., 2003. At Risk: Natural Hazards, People's Vulnerability and Disasters. Routledge, p. 496.

Wright, J., Ding, Y., 2016. Pathophysiological effects of particulate matter air pollution on the central nervous system. Environ. Dis. 1, 85–89. http://www.environmentmed.org/text.asp?2016/1/3/85/191932.

Xiao-Li, S., Yu, P., Hose, G.C., Jian, C., Feng-Xiang, L., 2006. Spider webs as indicators of heavy metal pollution in air. Bull. Environ. Contam. Toxicol. 76 (2), 271–277. https://doi.org/10.1007/s00128-006-0917-y.

CHAPTER FOURTEEN

Effects of climate change and natural disasters: Microevolution of spiders with aggressive behavioral traits

Abstract

The transitional and fragile ecosystems such as coastal ecosystems are too sensitive to maintain the state of dynamic equilibrium, being impacted by the actions of climate changes caused by tropical thunderstorms, torrential floods, torrential rainfalls, etc. Ecologists have observed that natural disasters affect the maintenance of inter- and intraspecific ecological dynamics among ecospecies of coastal and estuarine ecosystems, as they disrupt two cardinal factors, the food and habitat of ecospecies within the concerned ecosystems. Furthermore, scientists have witnessed that intermittent tropical thunderstorms (mainly hurricanes) within the coastal regions of the southeastern United States and adjacent areas have caused irreversible biobehavioral changes in invertebrate organisms, particularly spider communities in these ecoregions, and have indicated further that climate change may play a cardinal role in enhancing microevolutionary changes over time.

Keywords: Climate changes; Coastal ecosystem; Hurricane; Microevolution; Southeastern United States; Spiders.

As sea levels rise, the incidence of tropical storms will only increase. Now more than ever we must contend with what the ecological and evolutionary impacts of these storms will be for nonhuman animals.... Tropical cyclones likely impact both of these stressors by altering the numbers of flying prey and increasing sun exposure from a more open canopy layer.... Aggressiveness [of spider colonies] is passed down through generations in these colonies, from parent to daughter, and is a major factor in their survival and ability to reproduce.

Jonathan Pruitt

In the present climate change regime named Anthropocene, the environmental conditions of each ecosystem are gradually changing, and some fragile ecosystems such as coastal ecosystems are too sensitive to maintain

Spider Evolution
ISBN: 978-0-323-90041-6
https://doi.org/10.1016/B978-0-323-90041-6.00014-0

their state of dynamic equilibrium after impacts from frequent natural disasters such as monstrous tropical storms and hurricanes that can cause a cascade of immediate to long-term ecological responses. Many wildlife species inhabit ecological niches in estuarine and coastal habitats, and the cascading actions of tropical storms and hurricanes, comprising monstrous surges of storms, torrential rainfalls, wind vortices, enormous wind velocities, and extensive flooding, cause severe impacts on the maintenance of intraspecific and interspecific ecological dynamics in estuarine and coastal habitats. Hurricanes are natural disasters that redefine two essential conditions for a species—habitat and food—in the concerned ecosystem and can result in a biological species remaining alive as a part of short-term survival measures and a long-term evolutionary journey of diversification over time. The coastal ecosystem is considered one of the most suitable ecosystems (biologically and ecologically) on Earth, and more than half of the world population is located within 200 km of the shoreline of different parts of the world (University of Rhode Island [Hurricanes: Science and Society], December 6, 2010). In addition to their effects on animals, the vulnerable, cascading actions of tropical storms wash away vast volumes of coastal sands and loosely stratified soil layers to reshape the coastal landscape as scientists noticed that in just a 4-year interval, the sequential hurricanes Ivan in 2005, Katrina and Rita in 2005, and Gustav and Ike in 2008 altered around 100 miles of the topographical profile of the coastal shoreline of United States. It is further estimated that Hurricanes Katrina and Rita have reshaped a large part of the coastal shoreline by eroding the coastal landline of southeastern United States around 73 square miles as a combined action, caused by the monstrous wind blows with a cascade of dangerous waves.

Ecologists have observed that the barrier islands (sandy islands) have functioned as the potential storm barrier, are separated from the mainland by bodies of water such as bays, estuaries, and lagoons. Barrier islands and their adjacent buffering wetlands exposed to hurricane vortices lose substantial land area and its buffering to protect wetlands (the important ecological hotspots of species diversity) in the long run. The quantitative estimation of the recurrent erosion of barrier islands due to yearly whiplashing actions of hurricanes and anthropogenic actions, Louisiana barrier island in the southeastern United States, the shoreline is receding at a rate of 20 m per year along with the depletion of wetlands of 40 km^2 per year. If the cascading action of hurricanes remains uninterrupted, the pristine wetlands adjacent to the Louisiana barrier islands might have been disappeared in the next few years. The Louisiana barrier islands would have disappeared within a few decades (University of Rhode Island [Hurricanes: Science and Society],

December 6, 2010). In the preliminary assessment report of the storm-induced destruction, made by the post-Hurricane Hugo in 1989, around 1.8 × 106 ha of the coastal forests of South Carolina, USA, were observed to have been severely damaged due to combined actions of wind vortices and water waves (Bainbridge, 7th September, 2017). Similarly, Hurricane Andrew in 1992 had affected a large part of the mangrove forest area (forest of the Everglades) in Florida, damaging 100% of the mangrove trees and damaging the canopy of most trees, leading to collateral damage to a vast number of epiphytes and many animals and birds inhabiting the canopy areas of the mangroves (University of Rhode Island [Hurricanes: Science and Society], December 6, 2010).

The destruction of the canopy areas of coastal forest regions leads to the direct exposition of the forest floor with hyper-level beams of light, the high extent of wind exposition, and higher susceptibility to desiccation, leading to abrupt ecological alteration under storied forest cover. Hence, the passage of a hurricane not only tilts the ecosystem dynamics of the arboreal ecosystems by changing the inter- and intraspecific relationships of flora and fauna at the canopy level but also has enormous implications in the trophic dynamics of under storied forest ecosystems by changing the species diversities of crickets, insects, spiders, lizards, snakes, peafowls, rabbits, etc. as hurricanes in a core forest area would most often change the availability of the food and the mode of predation in that forest ecosystems. But the matter of fact is that each animal species would react in such a postcatastrophic regime in its way as ecologists, population biologists, and ornithologists have critically observed and analyzed that Hurricane Hugo hit the vast areas of old-growth forest in Carolinas in the United States in 1989 and it destroyed most nesting trees (which was around 87%) of the Red-cockaded woodpecker (*Picoides borealis*), resulting in immediate declination of 67% of the population of this endangered species. Conversely, after Hurricane Hugo landed in El Verde, Costa Rica, the frog population (globally diminishing amphibians) surged to six times more than its prehurricane level (University of Rhode Island [Hurricanes: Science and Society], December 6, 2010).

In bat species, the population varies from species to species. As population biologists observed that the Jamaican fruit bat species (*Artibeus jamaicensis*) and Red fig bat species (*Stenoderma rufum*) had left their natural forest habitat at El Verde, Puerto Rico, migrated to the adjacent forest areas, and did not return until 2 to 5 years later, respectively, to recolonize mainly due to lack of shelter, food, i.e., fruits and flowers, that supply them nectar as well (University of Rhode Island [Hurricanes: Science and Society], December 6, 2010). On the other side of the story, there was a surge in the population observed among Antillean long-tongued bat species (*Monophyllus redmani*) after the passage of Hurricane Hugo as scientists had contemplated that the opening of the forest cover most likely rendered regeneration of certain trees that supposedly provide fruits and flowers (the source of nectars), helped the sudden population surge of it in that forest areas. According to several quantitative estimations, quantitative analysis, and impact assessment reports of posthurricane phase on the threatened wildlife resources of any fragile ecosystems performed by conservation and population biologists, the critically endangered population of the Cozumel thrasher (*Toxostoma guttatum*), once found only in Mexico's Isla Cozumel, had been pushed into extinction due to the monstrous effects of Hurricane Gilbert in 1988 (University of Rhode Island [Hurricanes: Science and Society], December 6, 2010). This documentation has further elucidated that the population of the endangered avian species of Puerto Rican Parrot (*Amazona vittata*) has drastically been reduced to half to its original size after the passage of Hurricane Hugo in 1989 (Bainbridge, 7th September, 2017).

Scientists have observed that thousands of migratory birds have inhabited near the coastal habitat of the Gulf of Mexico to move between the breeding ground in North America and the wintering ground in South and Central America. The migratory birds have utilized the Gulf of Mexico as a primary stop-over for replenishing their lost energy (fuel) and shelter, but it has been observed by scientists that passage of hurricanes would directly or indirectly alter the annual migration of the birds and that would shrink the population of migratory bird species as the severity of the storms would jeopardize the survival of the migratory birds as the exhausted birds failed to use the stop-over, for refuel and resting (University of Rhode Island [Hurricanes: Science and Society], December 6, 2010). Apart from migratory birds, the decreasing size of the wild bird populations after the passage of hurricanes is a combination of factors like storm-induced mortality along with the factor like storm-induced forceful dispersing movements that drive the birds from their nesting territories. According to the conventional perception, the extreme level of wind and rain enhances the mortality rate of an enormous number of birds in hurricane-prone zones, but this is not the only cause of decreases in the size of the bird population. Eight months after Hurricane Hugo in 1989, scientists engaged in the bird count noticed a substantial loss of bird (mostly comprising insectivores or nectar-eating birds and frugivores or fruit-eating birds) populations in the US Virgin Islands

(Bainbridge, 7th September, 2017). Population biologists and ornithologists noticed that 6–10 months after the hurricane, missing bird populations returned and recolonized their earlier habitats. This helped scientists theorize that combined factors like scarcity of food and shelter and disturbance from high-velocity wind and rain forced birds to move from their original habitats and return at the beginning of the next breeding season. Scientists thus hypothesized that storm-induced bird movements, rather than mortality, are responsible for decreasing bird populations in hurricane-prone zones (University of Rhode Island [Hurricanes: Science and Society], December 6, 2010).

However, the recurrent surge of the monstrous tropical storms and hurricanes in short intervals have scientifically been hypothesized beyond the scope of metrological observations, analysis, predictions, and forecasting protocol, rather these kinds of unpredictable, unprecedented natural catastrophes (such as cascading effects of hurricanes) have been considered "Black Swan" weather events (Drye, December 4, 2012). In this regard, it must be mentioned that Nassim Nicholas Taleb, the eminent finance professor, former Wall Street trader, and writer, wrote a book on credit crisis issues before the 2008 financial meltdown or global recession titled the *Black Swan*. The take-home message of the black swan hypothesis elucidates that a system supported and protected from risk is ultimately subjected to more risk of catastrophe in the face of unprecedented situations, and this characteristic is equally applicable in natural science. However, the experts in the field of atmospheric science observed that in this current regime of Anthropocene, the climates in our biosphere have dramatically been changed due to anthropogenic activities. The unprecedented changes in carbon sinks to carbon sources alleviated the level of greenhouse gases in our atmosphere that trapped the radiant heat energies in this biosphere and gradually heated our climate, leading to tropical storms and hurricanes more intense and that has many adverse side effects on the diversified fauna in our biosphere (Taleb, 2010; Drye, December 4, 2012).

As long as the impact of the natural disasters, specifically the tropical storms and hurricanes, are concerned, the environmental impacts of such storms on the bigger-sized vertebrate animals such as fishes, avifauna, reptiles, amphibians, mammals, etc. have been noted from time to time, but the studies about its impact on lower animals and invertebrates such as insects, spiders have not been prioritized by ecologists, environmental and evolutionary biologists. In the contemporary investigations on the impact of Black Swan weather effects, tropical storms, and hurricanes, the environmental biologists observed that the exposition of such climatic deadly turbulence cascading actions of rain, wind effect of hurricane enhanced the level of mortality of most social spiders, leaving a small part of the behaviorally aggressive ones that could sustain such survival challenges; eventually these aggressive traits have been transcended to following generations in the course of its evolutionary journey of life (McMaster University, August 20, 2019). To elucidate it further, Jonathon Pruitt, an evolutionary biologist in the Department of Psychology, Neuroscience and Behavior, McMaster University, Canada, said: *"It is tremendously important to understand the environmental impacts of these 'black swan' weather events on evolution and natural selection."* (Little et al., 2019; McMaster University, August 20, 2019). As the vortex of monstrous winds of hurricanes uprooted the trees, tip off the canopies, defoliate the trees, leaving the pile of debris scattered everywhere, the cascading action of the storms that appeared itself as a stressful environment after a climatic disaster posed a selective pressure on several organisms. So evolutionary biologists have opined that there should be a better clutter-free elucidation to understand the environmental impacts of the suddenly surged "Back swan" type of natural disasters that would streamline the evolution and natural selection process of any biological organism. Jonathan Pruitt, the eminent evolutionary biologist, has further explained that in the regime of Anthropocene, when global climatic issues like global enhancement of greenhouse gases in our atmosphere induced gradual increasing of the temperature of our atmospheric environment, melting down of polar ice caps, raising the level of sea induced submergence of a large part of coastal plains, etc. directly or indirectly played the synergistic role in increasing the magnitude of the catastrophe of tropical storms and has indirectly impacted the ecological profile and evolutionary journey of any biological organisms, specifically the invertebrate organisms (McMaster University, August 20, 2019).

To elucidate the evolutionary impact of tropical storms and hurricanes on animals, evolutionary biologists have picked a social spider species, *Anelosimus studiosus*, which has lived in the colonies in the 3-dimensional tangled webs, the overhung structures over bodies of water such as rivers, lakes, and so on, and this spider inhabits a vast stretch of North America and South America including the Gulf and East Coasts of the southeastern United States and Mexico. The entire geographical regions of the Atlantic coast, inhabited by *A. studiosus*, are wrecked by the cyclonic turbulence at different times between May and November (Cross, August 22, 2019; McMaster University, August 20, 2019). The critical observation of the arachnologists and evolutionary biologists revealed that more than 100 female members of *A. studiosus* have lived in each of these webs. The meticulous observation of the arachnologists and evolutionary biologists ascertained that there are two distinct ecotypes as well as behavioral phenotypes in the species of *A. studiosus* living in that Atlantic coastal plain (Cross, August 22, 2019; McMaster University, August 20, 2019):

1. Docile group: individuals are even-tempered, move slowly, and responded quietly to the turbulence of tropical storms cyclones in the Atlantic coasts in and
2. Aggressive group: individuals are aggressive, move fast, and behave in a hyperactive manner in the face of tropical storms.

Though the two groups share a colony and live "side by side," scientists noticed that the preponderance of the aggressive individuals in a particular colony rendered the typical attribution of such colony to recognize it as an aggressive colony. The careful observations on the behavioral phenotypes, done by population biologists on the vertical transmission of genetic materials as well as characteristic features of *A. studiosus* from one generation to the next, deciphered that aggressive traits are heritable (Drye, December 4, 2012; McMaster University, August 20, 2019). The aggressive *A. studiosus* is a respondent to its predators and prey. The aggressive members of *A. studiosus* cannibalize other members, specifically males, and their own eggs and attack docile members, killing and devouring them along with other species of spiders and insects in their surroundings (McMaster University, August 20, 2019). So the environmental biologists have observed that cascading exposition of cyclonic storms in a particular habitat rendered the ecological and evolutionary profile of the concerned occupants of that habitat, which would transcend from one generation to the next generations. Such ecological and evolutionary changes would have created two possible scenarios, either promoting a particular section of behavioral phenotype over another and that might generate the divergence of unique species or species varieties as a result of evolutionary divergence of the existing phenotypic profile or the extinction of the weakly responding behavioral phenotype to the cascading turbulence of the tropical cyclones (Cross, August 22, 2019). The researchers have noticed that transmission of such behavioral phenotypic trait of "colony aggressiveness" has been transcended from parent's colony to daughter's colony that determines the survival and fecundity and which is integrated to the specific site or habitat, exposed to the recurrent tropical cyclones (McMaster University, August 20, 2019). The research observation of Pruitt and his research colleagues that has been published in the journal of Nature Ecology and Evolution, Pruitt said further, *"More broadly, aggressiveness in spiders is correlated with habitat resources, which also fluctuate in response to tropical cyclones"* (Little et al., 2019).

To contradict the observation of Jonathan Pruitt, Sergio Henriques, the Chair of IUCN's Spider and Scorpion Specialist Group, Zoological Society of London has observed that he contended that the contemporary findings of evolutionary biologists on the impact of climate changes such as the effect of tropical cyclones, which has rendered the social spiders more aggressive, received the unsurprising media attention that elicited fear in many people and causes negative views about this harmless creature. Sergio Henriques has mentioned in an undertone of sarcasm, *"... I have studied these arachnids for more than 15 years and I am not too concerned about tropical cyclones making them more aggressive. It is worth worrying about spiders themselves though. People who dislike spiders will, of course, be alarmed by any news of them getting more aggressive or even bigger due to climate change"* (Henriques, September 2, 2019). Refuting the observation of Pruitt (Little et al., 2019), Henriques contended that any local issue or issues might trigger the issues such as getting bigger for any species, and the studies of the team Pruitt have appeared to be localized in nature, coordinated on a target species rather than putting a global tag such as integrated to the global climate change issues that are less likely to be evidenced on the 48,359 species distributed worldwide (Henriques, September 2, 2019).

To criticize further the observation of Pruitt (McMaster University, August 20, 2019; Henriques, September 2, 2019) has tendered the reference of *Araneus diadematus*, or the Garden-orb-weaver, one of the largest common spiders in the Northern hemisphere that has reportedly attained its maturity at the end of the summer season, and that revealed that attaining to its maximum size is related to the change in seasonal cycles, rather than going through the cyclonic vortex and appeared to be impacted by ecological and evolutionary course of diversification in Spatio-temporal backdrop. Denouncing the observation of the researchers under the leadership of Pruitt, Henriques stated that the aggressiveness of any spider species must be defined by

1. the potential of some spider species to run with greater velocity to hunt prey versus other species that do not run aggressively for food if they are not hungry
2. spider species with a higher degree of prey-sharing efficiency and the lesser tendency of wasting food, such as when the concerned species

does not have enough appetite for hunting and does not want to waste leftover food

3. spiders with a tendency to cannibalize the males and eggs, and the choice between these options varies based on their demands
4. species with less susceptibility to being infiltrated by exotic spiders, e.g., when mingling with others is difficult, take a tough stand and do not share food or space

So according to the perception of aggressiveness of any species such as a spider, Henriques (September 2, 2019) has further contemplated that if the potential prey is not an insect, any vertebrate species such as a human, should not worry about the "aggressiveness" of the social spiders, except a few venomous spiders that possess a paralyzing immobilizing effect on the nerve and sensory systems of higher vertebrate prey species. According to Henriques (September 2, 2019), the size of any species has nothing to do with the aggressive behavioral trait, as exemplified by a carnivorous species such as a tiger, which is smaller but more aggressive than an elephant, a large herbivorous mammal.

However, the explanation tendered by Henriques (September 2, 2019) about gradual accumulation of behavioral phenotype as an aggressive colony in the regular colony of *A. studiosus* is not so satisfactory as he referred *A. studiosus* as a Communal spider, reported in individual states or groups of a few hundred individuals born to either a docile or an aggressive behavioral phenotype, which happens in only a few species as the number of Communal spiders are fairly low in respect to its global representation, but that does not justify the hypothesis that recurrent expositions along with intense and cascading actions of cyclones can alter the ratio between the docile and aggressive individuals in a spider colony of *A. studiosus*.

However, the research methodologies adopted by the team Pruitt were quite interesting to analyze the impact of storms that rendered the spider aggression. The researchers studied the storm's landfall path from the previous metrological records and followed the extreme weather alert and cyclone alerts to predict the likelihood landfall path of the incoming cyclones and examined the spider sample colonies in that particular location (Little et al., 2019). The same research team has also done the sample analysis 48 h after the storm passed the same locations. In their field surveys in 2018, researchers have specifically chosen three major cyclone events and sampled 240 spider colonies randomly. It was not an easy process as it involves logistical and methodological challenges like grasping the idea of the trajectory of the incoming cyclones based solely on the combination of anticipation and analysis of the previous records (Cross, August 22, 2019). However, during their sample survey, the members of team Pruitt have noted down the number of egg cases in each colony and the survival rate of spiderlings in it. The research colleagues, under the leadership of Pruitt (Little et al., 2019) have observed that the survival rate of the spiderlings in the spider colonies, after the first strike of the cyclone, was fairly high (which was around 75%) but the rate of survival of spiderlings and the number of egg cases of the spider inhabiting those colonies, started declining in the next and following strike of cyclones (Little et al., 2019). The researcher mentioned their critical observations in their research documentation *"tropical cyclones selected for more aggressive colony phenotypes. Following tropical cyclones, colonies with more aggressive foraging responses produced more egg cases and had more spiderlings survive into early winter, whereas the opposite trend emerged in control sites"* (McMaster University, August 20, 2019). The research team of Pruitt has further observed *"This trend is consistent across multiple storms that varied in size, duration, and intensity. This shows that these effects are not idiosyncratic but are robust evolutionary responses that hold across storms and at sites occupying a spread of 5-degree latitude"* (McMaster University, August 20, 2019). Though scientists were not in a position to define it clearly but contemplated that it must be triggered due to the shortage of food in and around their colony after the passage of storms and the researchers have assumed further that it seems to be an interspecific struggle to reach limited resources such as food, as competitors seem to be more aggressive to control limited food resources, thus causing a sect of spiders in the colony to respond behaviorally aggressively to protect the colony from foraging attacks by rival groups of invading spiders (Little et al., 2019). Researchers have observed that aggressive spiders are in a better position to acquire food resources in the potential scarcity of food resources, but this acts as a survival backlash because the aggressive members engage in infighting with either behavioral phenotypic counterparts (such as fighting with other aggressive members of the same colony) or different behavioral phenotypes (such as aggressive versus docile members of the same colony) when colonies are overheated (Cross, August 22, 2019). Pruitt explained further that *"tropical cyclones likely impact both of these stressors by altering the numbers of flying prey and increasing sun exposure from a more open canopy layer"* (Little et al., 2019). At the end of the experiment, researchers felt that there was a valid question, which had been raised by some members of team Pruitt, whether there was any factor that could intrigue the young spiders to be behaviorally aggressive instead of their adult mothers. It has been hypothesized further that the adult mother spiders were most likely engaged in finding and protecting limited food resources, as recurrent turbulence tends to minimize it after the passage of each cyclonic vortex, and they were tied to maternal care, thus forcing out their offspring at an early stage to become acquainted with survival skills, which requires the acquisition of behavioral aggressiveness to survive. These field-based experimental observations suggested further that after the passage of a tropical cyclone event, the spider colonies with more aggressive foraging responses have produced an increasing number of egg cases that yield a large number of spiderlings that survived the beginning of the winter season on the Atlantic coast of the United States and Mexico (Cross, August 22, 2019). The research scientists under the leadership of Jonathan Pruitt further contended that this trend was consistent irrespective of the extent of the vortex, duration, and intensity of the tropical storms, and that kind of consistency further indicated that tropical cyclonic storms or hurricanes have a massive evolutionary impact on no-human animals, specifically the small-sized invertebrates like spiders to render them sustained in the diversified form in Spatio-temporal course of evolution (Little et al., 2019).

Let's dig deeper to find out how changes in the physical or social environment, influenced by global climate change, create challenges to the anatomical setup and its functional mode of action in the brain of the spider species. The environmental biologists and neurobiologists revealed that the brain, the key regulator of the animals' central nervous system, is the key modulator responsible for the change in behavior in response to environmental cues. So to understand the response to the changes in environmental cues about the neurobehavioral traits of any animal, observation and analysis of anatomical changes in neurophysiological and anatomical studies of the brain of the concerned animal are inevitable. Based on their observation and analysis, scientists have contemplated further that *"interactions with conspecifics require higher brain complexity compared with interactions with the physical environment"* (Steinhoff et al., November 22, 2017). To find out further whether "early environmental conditions" have any effect on the development of the brain of spider species, and if so, which factors play cardinal roles, scientists have chosen the Jumping spider, *Marpissa muscosa,* as the model organism collected from distinct social and physical conditions (Steinhoff et al., November 22, 2017). The main experimental objective of such an experiment was to determine whether the factor like living in a complex or simple environment would make a difference in the neuroanatomical profile of the jumping spiders. Scientists are curious to understand whether growing up with or without conspecific would make any difference in the structural development of the brain of these jumping spiders, which might have been witnessed in their behavioral traits. The existing experimental observations intrigued scientists and arachnologists to consider Jumping spider as an exceptional arthropod, which has a wide array of complex behaviors, often witnessed in its "intense hunting behavior" and its extraordinary, "multimodal courtship" behavior. The animal biologists and biobehavioral experts contended that the jumping spiders are behaviorally competent enough to learn and perform complicated tasks. A careful comparative investigation on the optical and visual efficiency of neurosensory organs of the invertebrate and smaller size mammals ascertained that the eyesight of the jumping spiders is among the best in arthropods, and comparatively better than some mammalian species. From the earlier studies, scientists have contemplated that the Jumping spider is competent in "reversal learning[1]" and subjected to respond differently regarding behavioral and learning potential under varied habitat situations (Steinhoff et al., November 22, 2017). To find out whether the earlier environmental conditions would affect the volume of two higher-order brain centers (viz. mushroom body, or MB, and arcuate body, or AB[2]) of jumping spiders, Philip Steinhoff and three research colleagues carried out neurological experiments of the central nervous system (CNS) of the Jumping spider, which is plastic and the structure of which is altered according to the nature of external stimuli on it (Steinhoff et al., 2018). The comparative studies of MB and AB have been done by analyzing the micro-CT images. In their experiments, Steinhoff et al. (2018) have reared the female jumping spiders in three environments:

[1] Reversal Learning: In a particular situation when any animal, specifically a human being, is tutored to respond exponentially to two stimuli (which is determined from the behavioral response of the concerned animal from its response of approach or avoidance) under the binary results at the end of learning in the form of reward or punishment, succeed by giving another training under reversed reward systems (i.e., the approach which was previously rewarded is punished and the previously punished is rewarded).MB and AB: The MB and AB of the two higher-order brain centers of jumping spiders are considered responsible for visual image processing in the brain of jumping spiders, but the AB part has been assumed to play an important role in the control of locomotion.

[2] MB and AB: The MB and AB of the two higher-order brain centers of jumping spiders are considered responsible for visual image processing in the brain of jumping spiders, but the AB part has been assumed to play an important role in the control of its locomotion. MB and AB: The MB and AB of the two higher-order brain centers of jumping spiders are considered responsible for visual image processing in the brain of jumping spiders, but the AB part has been assumed to play an important role in the control of locomotion.

1. solidary or D (deprived condition),
2. solitary but physically enriched environment (P), and
3. mixed but physically enriched environment (together with other siblings).

In addition, wild species collected from natural habitats have been separately studied. Scientists in this experiment hypothesized that spiders reared in P and G should have higher brain volumes (bigger volumes of MB and AB) than D, as the physically enriched environment is cognitively assumed to be more demanding than the physically deprived environment. Scientists observed that spiders of the same species reared in physically enriched conditions had enlarged AB compared to counterparts, reared in nutrients deprived conditions, and spiders from nutrient-rich environments matured with better exploration potential than their counterparts (Steinhoff et al., November 22, 2017). Steinhoff and colleagues found that spiders in P environments have the largest brain volume compared with spiders reared in D environments, and spiders reared in G have a transitional size of the brain volume between that of D and P (Steinhoff et al., 2018). Scientists critically observed that the AB region of CNS is bigger than the MB region when a comparison between P and D was compared, and AB is the center of locomotor in the brain of spiders. Separate comparative investigations of the AB region between laboratory-reared spiders and spiders, collected from wild habitat (W), have revealed that the AB region, as well as the development of the brain of the jumping spiders in wild habitat, were much smaller, supposedly due to the scarcity of nutrient enrichment or limitation of food. So the experimental observation of the team Steinhoff revealed that the growth of the absolute brain volume and the relative brain volume of the higher-order brain centers plastically impacted the alteration of environmental conditions (Steinhoff et al., 2018). This experimental observation of Steinhoff ascertained that environmental conditions played a critical role in the developmental process of the brain of spiders. The unique demonstration of the brain development process of jumping spiders concerning environmental changes proved further that neuroplasticity is not restricted to only higher organisms or vertebrates and advanced nonvertebrates but also tiny arthropods like spiders (Steinhoff et al., November 22, 2017), and this neuroplasticity would help these creatures acclimatize to changing environmental conditions, which might be regarded as neurosensory adaptations, which is contemplated to play a cardinal role to guide the concerned organism down the trail of the evolutionary journey of life. It could be extrapolated further from such experimental observation that the cascading actions of frequent natural disasters, cyclonic tropical storms, or hurricanes would inevitably create a scarcity of food or nutrient deficient situations, which supposedly affect the brain developing process and CNS in spider species and alteration of environmental conditions might trigger the functional alteration of "center of locomotor control" or AB region of the brain of spiders, supposedly resulting in the alteration of behavioral traits of spiders in the regime of global climate changes.

The impact of climate change issues on spiders has been elucidated by Henriques (September 2, 2019) in a different way when he tendered a different example of the impact of climate change issues on the fecundity of *Eresus sandaliatus* or Ladybird spider (Henriques, September 2, 2019). Referring to his contemporary researches on the reproductive cycle of the Ladybird spider in an undisclosed location (as he does not want to disclose the location of this spider species, to avert the chance of high rate of exploitation of wild germplasm, which has usually been done for promoting illegal pet trade, of this unique spider species for ensuring its healthy population size) Henriques disclosed his critical observation that the male species of this Ladybird spider is endemic to some parts of the Western Asian highlands. Henriques (September 2, 2019) has noticed that the male members of the Ladybird spider species, which are usually cued by environmental changes such as enhanced temperature levels, sexually matured in the unusual hot regime of the winter season and left their nest sites in search of their potential mating female partners. Female members of this Ladybird spider (*E. sandaliatus*) normally attain sexual maturity after obtaining enough food rather than attaining sexual maturity after environmental cues like their male counterparts do. Unfortunately, the early exit ventures for courtship with females of the male members of the male ladybird spiders have remained unsuccessful as it was too early for females to attain sexual maturity in the early winter season, which resulted in the death of most male ladybird spiders, left their nests much earlier than a due time of the seasonal cycle of the reproduction calendar; analogically, which has nicely been mentioned by Henriques (September 2, 2020), *"Like Romeo, these males died without their Juliet."* Unfortunately, it was a scary impact of the global climate changes on a spider species that could have pushed a vulnerable spider species to the brink of extinction in this contemporary regime of Anthropocene. Undoubtedly, it was a convincing presentation by Henriques (September 2, 2019) about the impact of global climate changes that would affect the sustainability of spiders itself rather than jeopardizing the sustainability of human species by aggressive groups of spiders affected by altered climatic factors.

Let's shift our focus to the hard-hitting ground of contemporary global warming hotbed, the polar Arctic, which is becoming hotter at a rate twice as fast as the rest of the Earth, mainly due to anthropogenic action of intermittent releasing of the heat-trapping, greenhouse gases in our atmosphere (Machemer, July 23, 2018). The gradual increase in the atmospheric temperature rendered a faster increase in heat in the Arctic, cascading the meltdown of polar ice caps. Scientists have observed that the meltdown of the permafrost, the frozen layer of soil, and the dead organic material on it resulted in cascading the detritivore fungi bacteria to decompose it and it further cascading the formation and releasing of greenhouse gases such as carbon dioxide and methane, the heat-trapping gases in the Arctic microclimate, expediting the climate change altered environment of Arctic region as it added to the high level of greenhouse gases in the atmosphere. Very recently, Amanda Koltz, Arctic Ecologist at Washington University at St. Louis, USA has published her work, in which she revealed that in the backdrop of global warming-induced climate changes in the Arctic, abruptly coordinate direct changes in population densities of Arctic wolf spiders, their eating habits, etc. and it begins an "ecosystem-wide cascade," that ultimately regulate the pace of decomposition of melting permafrost in Arctic region (Machemer, July 23, 2018). Klotz's research publication seemed to narrate the impact of the warming climate, which affects prey-predator relations in the Arctic region, but it gave a broad perspective that explains how changes in such relationships would induce the function of the broader ecosystem. The scientific investigations since 2009, the arachnologists and the environmental biologists have noticed that climate change impacted the population dynamics of the spider communities and specifically observed that gradually heated-up Arctic, where the living beings are exposed to early springs and long duration of summer, supposedly make the invertebrate creatures larger (e.g., Wolf spider) and larger-sized species can produce more offsprings in this warmer Arctic environment, which has been normally witnessed in cases of spider species living in domestic environments rather than their smaller counterparts inhabiting wild habitats (Machemer, July 23, 2018). Typically, the wolf spiders eat mostly small-sized insects and smaller-sized spiders of other species, but their favorite food is a smaller-sized fungus-eating arthropod called springtails. In densely populated colonies, the bigger-sized wolf spiders took part in cannibalistic predation and eat one another. So the primary ideas behind the drafting of the research outline, coordinated by Amanda Koltz and her teammates, the result of which has been published in the proceedings of the National

Academy of Science in the United States in 2018 (Koltz et al., 2018), were to find out whether the changing of foraging pattern of wolf spiders to eat more or less number of springtails would change the population dynamics of fungus population in the Arctic and the resulting impact of decomposition of Arctic fungus would further aggravate the microclimatic environment and alter the ecological dynamics of Arctic ecosystems or not. To coordinate her experimental investigations, Koltz and her teammates have set up a few 5-foot-long experimental stations in Alaskan Arctic regions and analyzed experimental data for two summers, and observed meticulously how the temperature and number of spiders changed the composition and equilibrium of the different organisms in those experimental stations, set up on the patches of decomposing permafrost matrix (Machemer, July 23, 2018).

At the time of their experimental observations, Koltz and her team members expected that whenever the experimental ecosystem stations became warmer, the wolf spiders would devour all springtails in the respective stations, but the entire team was stunned by the experimental results, as they noticed that in those experimental ecosystem stations, wherever there was a preponderance of spiders, they ate a smaller number of springtail preys. So a large number of wolf spiders in any station includes many springtails, and the large springtail population eats all of the fungi of that experimental station, pushing down the rate of decomposition as well as lesser carbon releasing from the carbon sink of the experimental permafrost stations, rather than stations with fewer spiders with a high rate of fungal decomposition (Machemer, July 23, 2018). In support of her experimental observation, Koltz said, "*The way the organisms interact with one another can influence ecosystem functions such as how much carbon stays fixed by plants, how quickly decomposition happens, and how quick the nutrients are cycled within that ecosystem ... controls on nutrient cycle in the Arctic are very important for us to understand because the region plays a disproportionately large role in the global carbon cycle*" (Koltz et al., 2018). So it was a miraculous finding that team Koltz observed the presence of wolf spiders or their potential population surge would help combat the negative impact of climate changes in the Arctic. Joseph Bowden, a renowned entomologist in Canadian Forest Service, has mentioned, "*The novelty of Dr. Koltz's paper is that it shows [climate change] not only having direct impacts on these important ground-dwelling animals but also on the complex ecological interaction between species on the tundra*" (Machemer, July 23, 2018). Does it necessarily mean that the preponderance of these tiny wolf spiders in Arctic permafrost would save us from the grime impact of climate changes when decomposition of organic material buried in the permafrost has been exposed as the permafrost started melting in this current warmer regime of Anthropocene, has no direct conclusive answer from realistic viewpoint? Yet Koltz firmly expressed her optimism as she said, *"this suggests that under some circumstances, they could be alleviating some of the effects of warming on carbon losses from the tundra. It is a good thing"* (Linsey, July 27, 2018).

The experimental observation of the climate change impact on wolf spiders is a ground-breaking experiment, yet several gray areas remained unexplored. As scientists were not clear about what made the wolf spiders with a high-density population lose the appetite that held them to stop eating springtails? What was the exact reason behind the act of foraging of Wolf spider, to shift its focus from springtails to engaged in cannibalism and eat the interspecific and intraspecific members of spiders, was not clear, and it seemed that perhaps its higher range of temperature or some internal physiological and molecular changes that altered the existing prey-predator dynamics?

Sustainability of any biological entity is not a self-determinant issue; rather, it is an integral part of interactions between that biological entity and the human beings, and that is integrated into its proximity of its cohabiting in the human ambiance. There is no doubt that the living of any organism in its anthropogenic vicinity must have been impacted by the human actions in this current age of Anthropocene, where micromanagement of every ecosystem has been decided and guided by anthropogenic actions, ardently practiced since the beginning of the Industrial Revolution. So in this current regime of anthropogenic actions triggered global climate changes, the sustainability of any species is directly or indirectly depends on the human impact on that particular species; it does not matter how large an animal is, such as a blue whale, or how small it is, such as a tiny arthropod spider. From the traditional folk tales to the contemporary toxicological studies, it was a wide array of publications that elucidated the negative impacts of spiders as a bad omen or fearsome creatures on human life except for a few exceptions in the glorified history of Indian American cultures and traditional African aboriginal histories, where the spider has been considered trickster god "a symbol of patience and endurance, an icon of intelligence and skillfulness to clan animals or even Native American Indian tribes with spider clans such as "Kookyangwngyam" or "Koking-wungwa" of the Hopi tribe (http://www.native-languages.org/legends-spider.htm, April 21, 2012)." In the traditional native Indian American culture, the spider has a glorified place for being recognized as a sacred entity and communicating role-plat to build a bridge between this real world to the world of surrealism, and it also helps protect traditional aboriginal life and ensure protection and conservation of nature and patronage of traditional and natural ways of organic living followed by the American Indian and African aboriginal communities in North America, South America, and various parts of Africa. On the other side, the human impact on the tiny arthropods, the small creatures such as spiders, has not been addressed as we ignored the ecological importance of this so-called "inconsequent nature of invertebrate" maintenance of the ecological equilibrium of nature in this current regime of global climate changes. However, two Spanish arachnologists, Samuel Prieto-Benitez and Marcos Mendez from FECYT-Spanish Foundation for Science and Technology, King Juan Carlos University, Madrid, Spain, have tried to find out whether spiders are more tolerant toward human impact than any other animal species or not and found that spiders suffer in the same way, human species have impacted the other animal species as Prieto-Benitez and Mendez (2011) have mentioned it in their research communication as they said: *"The abundance and number of spider species are negatively affected by the impact of many human lands uses, such as habitat fragmentation, fire, and pesticides."*

The poor presence of threatened spider species in the Global Red List of International Union of Conservation of Nature and Natural Resources (IUCN) was considered vogue by Prieto-Benitez and Mendez (2011), which made them undertake secondary research by undertaking a metastudy of 173 research papers and review articles, published since 1980, that helped them analyze, elucidate, and publish a generalizable data. Prieto-Benitez and Mendez (2011) have focused their studies on human impact on spiders inhabiting in the three ecosystems: farmland, pastureland, and woodland, and they observed that there are direct damaging effects of human operations (the mechanical operations along with biochemical applications) in farmlands and pasturelands made substantial damage of spider populations, inhabiting such ecosystems, but they were not very sure about the similar impact on spiders, living in woodland ecosystems, though they observed that habitat fragmentation in the woodlands caused a substantial negative impact on the spider population in the wild habitat. Prieto-Benitez and Mendez (2011) have noticed that in the farmland and pasture ecosystems, all over the world, the anthropogenic operational issues cultivation of conventional crops, fires, sheep-grazing, etc. have extraordinary adverse effects on the arachnid fauna of the concerned ecosystems as it leads alteration of nature of wild vegetation in the ambiance of such anthropogenic operations.

The high level of mechanical operations in the farmland, plantation of a commercial cash crop such as rubber plantation or coffee plantation, and the extensive use of pastureland to promote cattle farming, even by deforesting vast areas of Amazonian forest land, must be managed sustainably to manage the ecological equilibrium of any fragile ecosystem, where the issues such as the existence of the small arthropods such as spiders are part of it. Along with the sustainable reduction in the current usage of the extensive level of mechanical operations and grazing operations, there should be a sustainable operational policy of limited or no use of harmful chemical insecticides and pesticides in the agricultural farmlands, orchards pastureland ecosystem to maintain the diversity and abundance of the spider communities, inhabiting in the concerning ecosystems. So as a feasible option, the idea of organic farming has been promoted to maintain the biodiversity and sanctity of our pristine nature to protect it from further devastation. In a recent experimental observation of Korento et al. (2020), the researchers have observed that neonicotinoids that act as potential insecticides in the farmlands to keep agricultural lands and farmlands free from agricultural pests that cause collateral damage to the biodiversity of the concerned agroecosystems by overkilling spider communities (e.g., *Pardosa agrestis*). In addition, many individuals die from exposure to and absorption of the aerosols of neonicotinoids inhabiting those forest ecosystems. Thus, the overuse and application of chemical insecticides in traditional agricultural practices exterminate harmful insects and crop-damaging pests and kill spider communities in agricultural ecosystems and farmlands. As a cascading of traditional farming with chemical insecticides and pesticides, it would hinder the principle of biological control of insects and pests of the concerned ecosystem and not only increase the overall cost of production of the crops but also induce collateral damages of the environment by infiltrating the toxic and harmful chemicals (some have potential mutagenic effects on the biological organism in molecular level) in a vast array of potential toxins, mutagens, and carcinogens that enter each biological organism through food trophic of that ecosystem's food web, and that has an insurmountable effect to achieve sustainability of any species in a long run whether the species would be a big mammal or tiny arachnid.

For paleontologists, it was difficult to determine the evolutionary history of spiders from the Stone Age 400 million years ago, as very few stones were laid down on Earth to act as a paleontological repository to narrate the progress of life. Likewise, it was an almost impossible task for molecular biologists to narrate the stepwise evolutionary progress of arachnids, read from their

DNA, and define the origin and evolution of spiders on Earth. As the diversification of arachnids happened so early, it did not leave enough trace of evolutionary milestones and mutational changes in their genome. Yet the prehistoric message of life, which has been encrypted on the stones and deciphered by evolutionary biologists, ultimately revealed that spiders, the successor of arachnids, were among the maiden land dwellers, which evolved around 420 million years ago on Earth (Pappas, March 31, 2016). Since the dawn of human civilization, in the days of early historical and mythological interactions of spiders with our predecessors (around 3060 years ago, David, the king of Jerusalem; or around 1400 years ago the prophet Mohammad, the founder of Islam), spiders appeared as saviors rather than a worrisome creature or a bad omen, mentioned in several literature and fictions. In fact, any living being (whether a big mammal such as a blue whale or a tiny arthropod-like ant) always goes through the trail of biological evolution and diversifies from its maiden appearance to ensure its sustainability.

As a long-living creature of the arachnid group, different species of spiders have gone through a reticulate trail of evolution, and some of the species have become extinct. In this current regime of climate changes, if the cascading impact of tropical storms and hurricanes pushed certain species of spiders or their behavioral phenotypes to enjoy better survival advantage, it would supposedly happen according to the rule of nature; there is no point to promote arachnophobia for this natural course of action, as spiders are helpful creatures in real life rather than the frightful creatures depicted in science fiction. So whether it is an issue such as biological control of insects and pests or combating climate changes in the tundra, we human beings need more help from the inconsequent, tiny creatures like spiders, rather than it needs any protective measure from us and that pointed out that conservation of population diversity of spiders is not a utopian idea but a hardcore reality to ensure the sustainability of *Homo sapiens* in the roller coaster metaphoric ride in this regime of Anthropocene. Most Native American Indian tribes consider the traditional handwoven Ojibwa (one of the leading Native American Indian tribes) craft, known as "Dream catchers," symbolic spiderwebs that filter out bad dreams, and most Native American Indians consider killing a spider to bring bad luck to the person who kills it (http://www.native-languages.org/legends-spider.htm, April 21, 2012). So in this turbulent regime of global climate changes, let us share the traditional principle of living of Native Indian Americans to protect spiders so they can spin spiderwebs, and those symbolic dream catchers might drive away our bad dreams of the possible catastrophic end of human civilization and protect us from the symbolic curse of the one-eyed god "Huracan." The great evolutionary biologist Amanda Koltz and her co-researchers expressed their concern to coordinate in-depth research specifically on the ecological and environmental role of spiders in mitigating the nefarious impact of climate change in the future, expressed her optimism as she said: *"We tend to forget about the tiny animals because they are not as visible as the bigger mammals .... it is really neat to think about how these small animals have important impacts on the ecosystem as well"* (Machemer, July 23, 2018).

References

Bainbridge, J., 7th September, 2017. In 1989, Hurricane Hugo thrashed South Carolina coast. https://www.greenvilleonline.com/story/news/local/greenville-roots/2017/09/07/hurricane-hugo-south-carolina-coast-1989/623180001/.

Cross, D.T., August 22, 2019. Climate Change Will Make Some Spiders More Bellicose. https://www.sustainability-times.com/environmental- protection/climate-change-will-make-some-spiders-more-bellicose/.

Drye, W., December 4, 2012. Powerful Hurricanes Such as Sandy and 'Black Swan' Storms Could Alter U.S. Coastline. National Geographic Society Newsroom. https://blog.nationalgeographic.org/2012/12/04/powerful- hurricanes-such-as-sandy-and-black-swan-storms-could-alter-u-s- coastline/.

Henriques, S., September 2, 2019. Spiders Are Threatened by Climate Change, Even the Biggest Arachnophobes Should Be Worried. https://www.firstpost.com/tech/science/spiders-are-threatened-by- climate-change-even-the-biggest-arachnophobes-should-be-worried- 7281651.html.

Koltz, A.M., Classen, A.T., Wright, J.P., 2018. Warming reverses top- down effects of predators on belowground ecosystem function in Arctic tundra. Proc. Natl. Acad. Sci. U. S. A. 115 (32), E7541–E7549. https://doi.org/10.1073/pnas.1808754115.

Korenko, S., Sýkora, J., Řezáč, M., Heneberg, P., 2020. Neonicotinoids suppress contact chemoreception in a common farmland spider. Sci. Rep. 10, 7019. https://doi.org/10.1038/s41598-020-63955-z.

Linsey, K., July 27, 2018. Climate Change Means Bigger Arctic Spiders — but Don't Worry, that Could Be a Good Thing. www.arctictoday.com/climate-change-means-bigger-arctic-spiders-dont-worry-thats-good-thing/.

Little, A.G., Fisher, D.N., Schoener, T.W., Pruitt, J.N., 2019. Population differences in aggression are shaped by tropical cyclone-induced selection. Nat. Ecol. & Evol. 3, 1294–1297. https://doi.org/10.1038/s41559-019-0951-x.

Machemer, T., July 23, 2018. Climate Change Makes Spiders Bigger—And That's a Good Thing. www.nationalgeographic.com/animals/2018/07/wolf-spider-arctic-climate-change-news/.

McMaster University, August 20, 2019. Hurricanes Drive the Evolution of More Aggressive Spiders. Science Daily. Retrieved August 30, 2020 from. www.sciencedaily.com/releases/2019/08/190820081833.htm.

Pappas, S., March 31, 2016. 305-Million-Year-Old 'Almost Spider' Unlocks Arachnid History. https://www.scientificamerican.com/article/305- million-year-old-almost-spider-unlocks-arachnid-history/.

Prieto-Benitez, S., Mendez, M., 2011. Effects of land management on the abundance and richness of spiders (Araneae): a meta-analysis. Biol. Conserv. 144, 683 691.

Steinhoff, P.O.M., Liedtke, J., Sombke, A., Schneider, J.M., Uhl, J., November 22, 2017. Early Environmental Conditions Affect the Volume of Higher-Order Brain Centers in a Jumping Spider. http://onlinelibrary.Wiley.com/doi/10.1111/jzo.12512/full.

Steinhoff, P.O.M., Liedtke, J., Sombke, A., Schneider, J.M., Uhl, J., 2018. Early environmental conditions affect the volume of higher-order brain centers in a jumping spider. J. Zool. 304 (3), 182–192.

Taleb, N.N., 2010. The Black Swan: The Impact of the Highly Improbable, second ed. Penguin, London, p. 444.

University of Rhode Island [Hurricanes: Science and Society], December 6, 2010. Ecosystem Perspective: What Can a Hurricane Do to the Environment? http://www.hurricanescience.org/society/impacts/environmental_impacts/.

Index

Note: 'Page numbers followed by "b" indicate boxes, those followed by "f" indicate figures and those followed by "t" indicate tables.'

D

E

F

MIX
Papier | Fördert gute Waldnutzung
FSC® C083411

Kolibri 360 GmbH
Ferdinand-Jühlke-Straße 7
99095 Erfurt, Deutschland
produktsicherheit@kolibri360.de